# α-TERTIARY AMINES EN ROUTE TO NATURAL PRODUCTS

VISUAL GUIDES TO NATURAL PRODUCT SYNTHESIS SERIES

# α-TERTIARY AMINES EN ROUTE TO NATURAL PRODUCTS

ABDUL HAMEED

MARIYA AL-RASHIDA

MUHAMMAD RAZA SHAH

ELSEVIER

Elsevier
Radarweg 29, PO Box 211, 1000 AE Amsterdam, Netherlands
The Boulevard, Langford Lane, Kidlington, Oxford OX5 1GB, United Kingdom
50 Hampshire Street, 5th Floor, Cambridge, MA 02139, United States



**Library of Congress Cataloging-in-Publication Data**
A catalog record for this book is available from the Library of Congress

**British Library Cataloguing-in-Publication Data**
A catalogue record for this book is available from the British Library

ISBN: 978-0-12-822262-1

For information on all Elsevier publications visit our website at
https://www.elsevier.com/books-and-journals

*Publisher:* Susan Dennis
*Acquisitions Editor:* Emily McCloskey
*Editorial Project Manager:* Tracy Tufaga
*Production Project Manager:* Sruthi Satheesh
*Cover Designer:* Miles Hitchen



Typeset by TNQ Technologies

# Contents

# Preface

To date, Nature continues to be the greatest reservoir of astonishingly diverse array of structural diversity observed in chemical space; interestingly a lot of this chemical space is speculated to be yet unexplored. All living organisms including the microbes synthesize a plethora of biologically active compounds de novo in their cells and tissues. These active molecules are vital for sustenance of life and associated delicate ecosystems. Since the beginning, it has been the deepest desire of man to harness this power of nature for therapeutic intervention and prevention of diseases. A significant chunk of modern medicinal and pharmaceutical chemistry is based on either pure natural products or synthetic compounds having scaffolds inspired and derived from the natural products. Hence, natural products, by virtue of their vast structural diversity, continue to hold the fascination of synthetic chemists.

Herein, we propose a book series that will cover the area of total synthesis of natural products. The multistep syntheses of natural products will be presented in easy-to-grasp schemes, highlighting the key steps involved in synthetic layout. This visual guide will provide a quick and easy way to read and to understand the new/novel synthetic strategies to construct the whole structural framework of natural products. The visual guide on natural products synthesis will deliver express access to read and understand the synthetic strategies. Volume I will cover the class of natural products bearing α-tertiary amine motif. Some of the examples of such natural products include altemicidin, amathaspiramide (A−F), kaitocephalin, lactacystin, salinosporamide, manzacidins, neooxazolomycin, sphingofungins, (1S,3R)-1-aminocyclopentane-1,3-diarboxylic acid (ACPD), and tetrodotoxin. These molecules have gained much popularity among synthetic chemists. A large number of synthetic approaches and total syntheses of these molecules have been published. Thus, the collection of natural product syntheses in this book will provide a quick and an easy access to the readers to get the gist of synthetic route toward the targeted natural products.

CHAPTER

# 1

# Natural products with α-tertiary amine

## 1.1 Abstract

Natural products isolated from various sources such as plants, marine life forms, fungus, algae, and bacteria have been the greatest source of design inspiration for synthesis of biologically active molecules and drugs targeting various diseases. These natural products provide an almost unlimited source of biologically active unique molecules and novel scaffolds. These molecules are constantly being synthesized de novo, inside the living cells, and serve various purposes from simple survival to growth and defense. Such molecules serve as a design template for synthetic chemists and medicinal chemists for the synthesis of molecules with specific biological activities. Most of the drugs available in the market are based on natural products. Alkaloids are a unique class of natural products that have been vital in the development of drugs against various diseases (Fig. 1.1).

It has long been the quest of synthetic organic chemistry to come up with the design and experiments for in-lab synthesis of naturally occurring molecules and their derivatives. Although the staggering structural complexity and multiple stereocenters in alkaloids have proved to be a challenge from synthetic chemistry point of view, this structural diversity is responsible for the activity of these molecules against various biological targets, hence their potential to treat different diseases.

On the basis of structural diversity, alkaloids have been classified into different groups, among which the natural products bearing α-tertiary amine moiety have attracted considerable attention from the synthetic community. Among the pool of natural products having *N*-quaternary center, we have selected the most popular natural products such as


https://doi.org/10.1016/B978-0-12-822262-1.00001-9

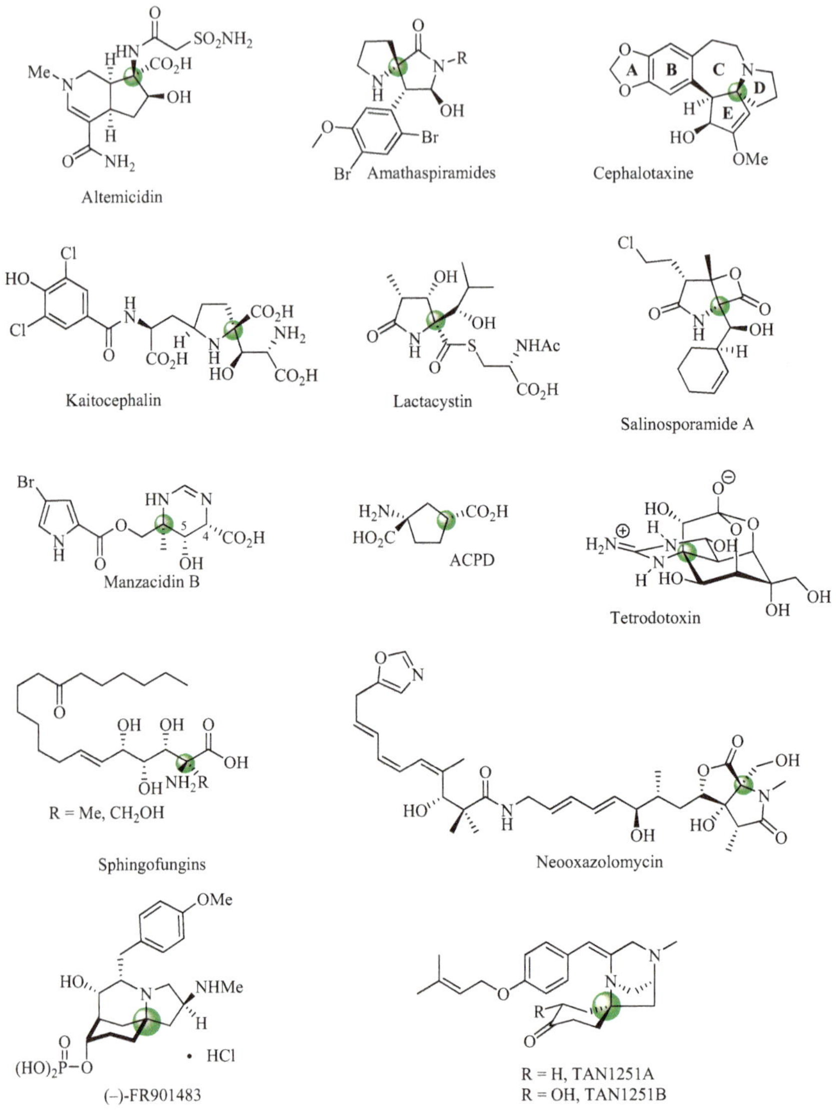


**FIGURE 1.1** Structures of different natural products containing *α-tert*-alkylamino carbon center.

altemicidin, amathaspiramides, cephalotaxine, kaitocephalin, lactacystin, salinosporamide, manzacidines, tetradoxtin, neooxazolomycin, sphingofungins, ACPD, TAN1251A & B, and (−)-FR901483 with diverse enantio- and diaseteroselective methodologies to construct tetra-substituted carbon center surrounded by three carbons and one nitrogen atom, i.e., $\alpha$-*tert*-alkylamino carbon center[1] or in case of bearing acid moiety, termed as $\alpha,\alpha$-disubstituted-$\alpha$-amino acid.[2]

## References


1. (a) Mailyan, A. K.; Eickhoff, J. A.; Minakova, A. S.; Gu, Z.; Lu, P.; Zakarian, A. Cutting-Edge and Time-Honored Strategies for Stereoselective Construction of C−N Bonds in Total Synthesis. *Chem. Rev.* **2016,** *116,* 4441−4557.
   (b) Kang, S. H.; Kang, S. Y.; Lee, H.-S.; Buglass, A. J. Total Synthesis of Natural *Tert*-Alkylamino Hydroxy Carboxylic Acids. *Chem. Rev.* **2005,** *105,* 4537−4558.
2. Ohfune, Y.; Shinada, T. Enantio and Diastereoselective Construction of $\alpha$, $\alpha$-disubstituted $\alpha$-amino Acids for the Synthesis of Biologically Active Compounds. *Eur. J. Org Chem.* **2005,** *2005,* 5127−5143.

CHAPTER

# 2

# Altemicidin

## 2.1 Abstract

(−)-Altemicidin is an interesting naturally occurring 6-azainden monoterpene sulfonamide alkaloid that exhibits strong inhibition against tumor cells and has potent acaricidal activity. (−)-Altemicidin **1** was isolated by Takahashi research group in 1989 from the actinomycete strain *Streptomyces sioyaensis*.[1] The acylated structure of the altemicidin showed strong inhibitory activity against isoleucyl, leucyl, and valyl *t*RNA synthetase.[2] The interesting biological properties and complex structural features of (−)-altemicidin make it an attractive target for synthetic organic chemists (Fig. 2).

The structural features of altemicidin include a core 6-azainden ring, *β*-hydroxy group at C-2 position, and key *α*,*α*-dialkyl-*α*-amino acid moiety at C-1 position. So far, only three synthetic approaches toward this alkaloid have been reported. The first enantioselective total synthesis of altemicidin was carried out by Kende's group via Diels−Alder reaction to construct C-1 *N*-bearing quaternary center and Potier−Polonovski rearrangement in the synthetic sequence. Later, Kan's group built the stereocontrolled core framework of altemicidin as an advanced intermediate. Further, the Kan's group competed the total synthesis of altemicidin derivative SB-203202 via desymmetrical C−H insertion reaction and Curtius rearrangement in an enantioselective way. The schematic outlook of all approaches has been presented in the coming section.

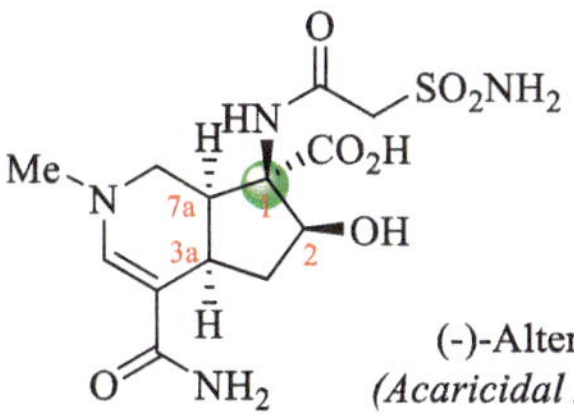


**FIGURE 2** Structure of altemicidin.


https://doi.org/10.1016/B978-0-12-822262-1.00002-0

## 2.2 Kende's first total (−)-altemicidin synthesis

See (Scheme 1).

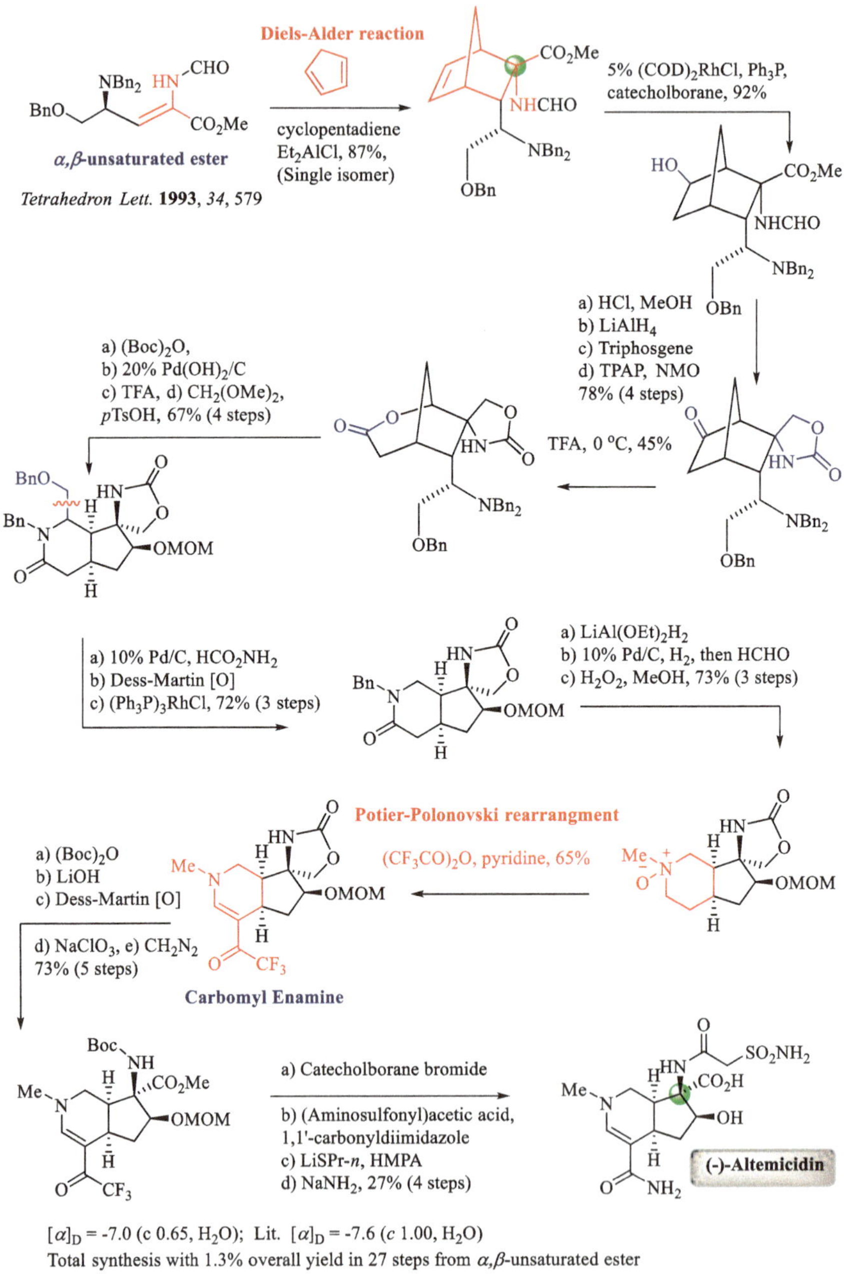


**SCHEME 1** Total synthesis of (−)-altemicidin via Potier–Polonovski rearrangement.[3]

## 2.3 Kan's approach toward altemicidin Bicyclo[3.3.0] framework (2008)

See (Scheme 2).

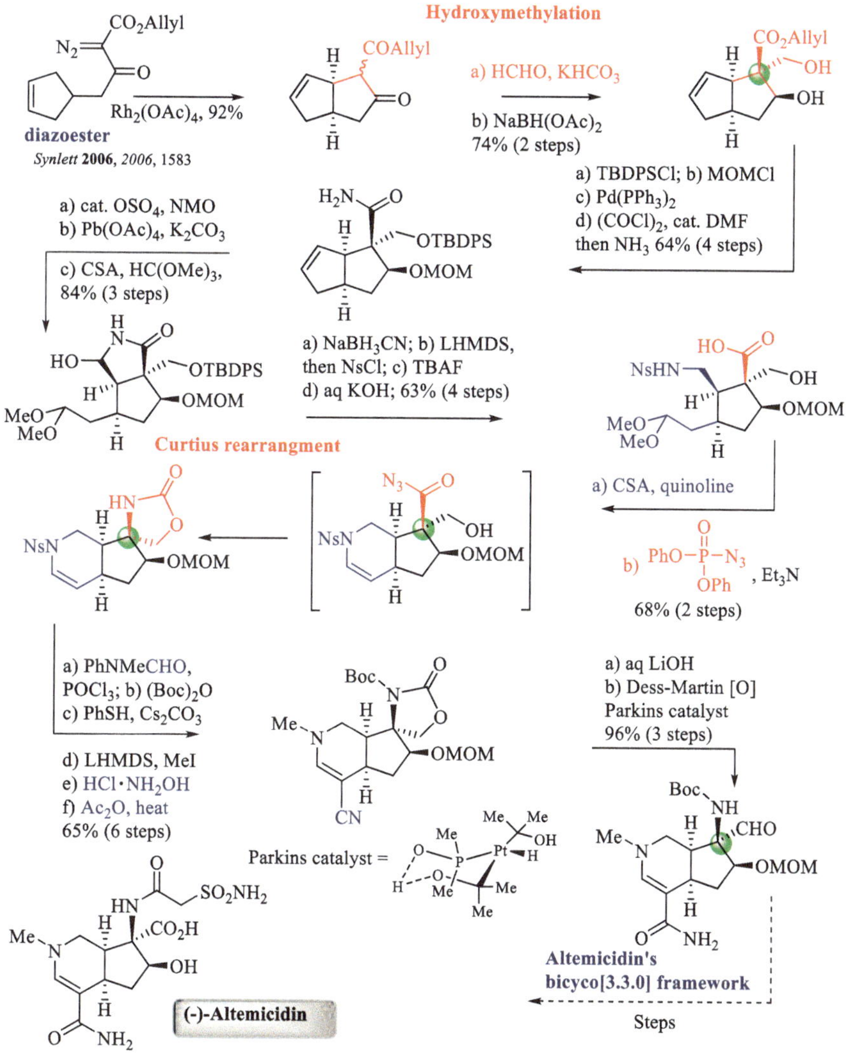


Synthesis of Altemicidin's bicyco[3.3.0] framework from diazoester with 9.4% overall yield in 27 steps from *α,β*-unsaturated ester

SCHEME 2 Stereocontrolled construction of altemicidin core framework.[4]

## 2.4 Kan's total synthesis of SB-203207: an altemicidin's analogue (2014)

See (Scheme 3).

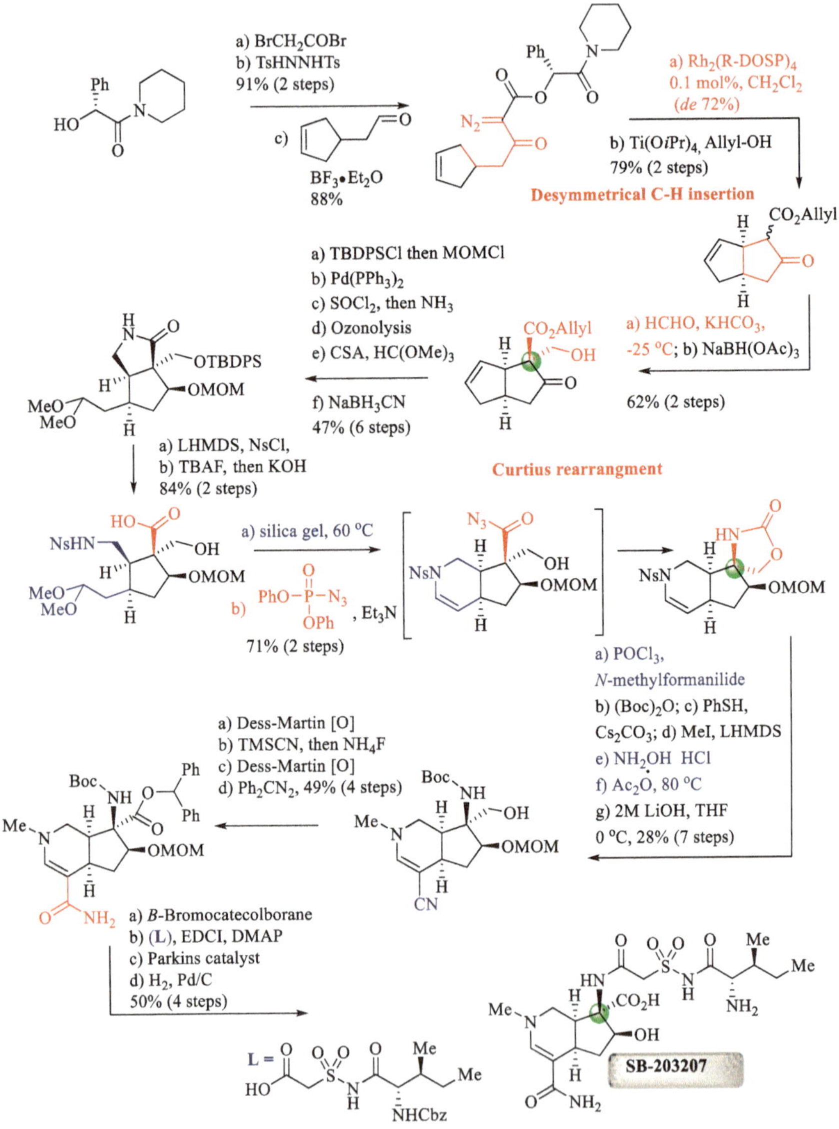


Total synthesis of SB-203207: An Altemicidin's analogue with 0.81% overall yield in 32 steps

$[\alpha]_D^{20}$ +25.9 (*c* 0.025, $CH_3OH$)

SCHEME 3 Total synthesis of altemicidin's analog (of SB-203207).[5]

## 2.5 Hayakawa's studies toward altemicidin's analogue (SB-203207)

See (Scheme 4).

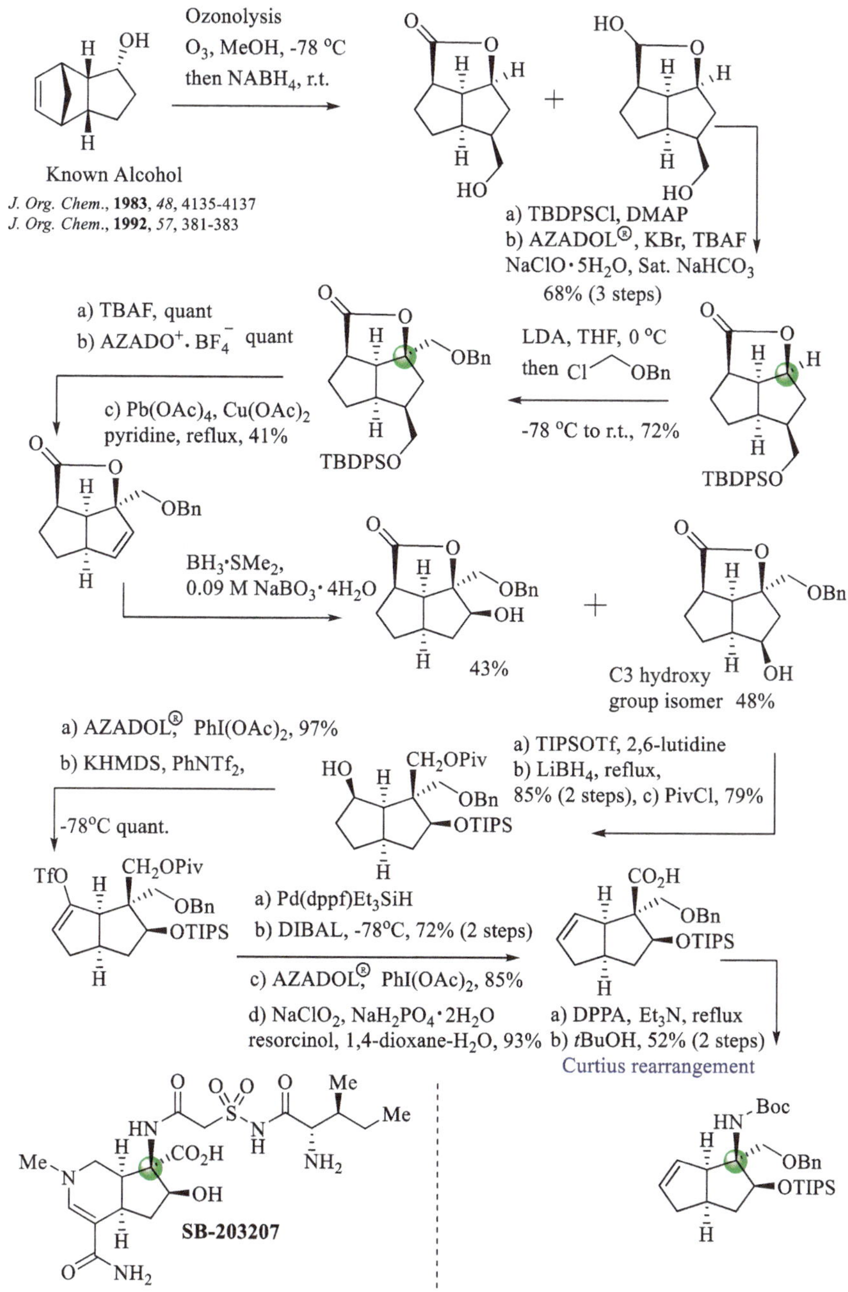


SCHEME 4 Altemicidin's analog: formation of four contiguous nitrogen-containing stereogenic centers.[6]

## References

1. Takahashi, A.; Kurasawa, S.; Ikeda, D.; Okami, Y.; Takeuchi, T. Altemicidin, a New Acaricidal and Antitumor Substance. *J. Antibiot.* **1989,** *42,* 1556–1561.
2. (a) Banwell, M. G.; Crasto, C. F.; Easton, C. J.; Forrest, A. K.; Karoli, T.; March, D. R.; Mensah, L.; Nairn, M. R.; O'Hanlon, P. J.; Oldham, M. D. Analogues of SB-203207 as Inhibitors of tRNA Synthetases. *Bioorg. Med. Chem. Lett* **2000,** *10,* 2263–2266.
   (b) Houge-Frydrych, C. S. V.; Gilpin, M. L.; Skett, P. W.; Tyler, J. W. SB-203207 and SB-203208, Two Novel Isoleucyl *t*RNA Synthetase Inhibitors from a Streptomyces Sp. *J. Antibiot.* **2000,** *53,* 364–372.
3. (a) Kende, A. S.; Liu, K.; Jos Brands, K. Total Synthesis of (-)-Altemicidin: A Novel Exploitation of the Potier-Polonovski Rearrangement. *J. Am. Chem. Soc.* **1995,** *117,* 10597–10598. and references therein.
   (b) Kende, A. S.; Brands, K.; Blass, B. A Novel Dyatropic Rearrangement of $\gamma$-*N,N*-dibenzylamino $\alpha,\beta$-dehydro *N*-Formylamino Acid Esters. *Tetrahedron Lett.* **1993,** *34,* 579–582.
4. Kan, T.; Kawamoto, Y.; Asakawa, T.; Furuta, T.; Fukuyama, T. Synthetic Studies on Altemicidin: Stereocontrolled Construction of the Core Framework. *Org. Lett.* **2008,** *10,* 169–171. and references therein.
   (i) Kan, T.; Inoue, T.; Kawamoto, Y.; Yonehara, M.; Fukuyama, T. A Novel Synthesis of Bicyclo [3.3.0] Octane Ring System via a Desymmetric CH Insertion Reaction. *Synlett* **2006,** 1583–1585, 2006.
5. Hirooka, Y.; Ikeuchi, K.; Kawamoto, Y.; Akao, Y.; Furuta, T.; Asakawa, T.; Inai, M.; Wakimoto, T.; Fukuyama, T.; Kan, T. Enantioselective Synthesis of SB-203207. *Org. Lett.* **2014,** *16,* 1646–1649.
6. Hayakawa, I.; Nagayasu, A.; Sakakura, A. Toward the Synthesis of SB-203207: Construction of Four Contiguous Nitrogen-Containing Stereogenic Centers. *J. Org. Chem.* **2019,** *84,* 15614–15623.

CHAPTER

# 3

# Amathaspiramides A–F

## 3.1 Abstract

The amathaspiramides A–F are a class of marine alkaloids that have been isolated from a New Zealand collection of the bryozoan *Amathia wilsoni* by Prinsep and Morris.[1] The core of these alkaloids is a highly functionalized aza-spirobicyclic structural framework possessing *α-tert*-alkylamino carbon center and a hemiaminal center. The amathaspiramides have exhibited excellent bioactivities as antiviral and antimicrobial agents, albeit with moderate cytotoxicity. This family of structurally complex natural products has attracted significant attention from the synthetic chemists. There are many different synthetic methodologies reported for total synthesis of different members of amathaspiramides A–F. Trauner and Ohfune completed the total synthesis of amathaspiramide F, whereas Fukuyama completed the total synthesis of A–F in a stereoselective way. The Tambar's group reported synthesis of amathaspiramide F, whereas Lee et al. completed total synthesis of amathaspiramide C; formal synthesis of A–F was reported via Fukuyama's route. Recently, Sun's group reported the synthesis of amathaspiramides B, D, and F in an asymmetric way (Fig. 3).

## 3.2 Trauner's first total synthesis of (−)-amathaspiramide F (2002)

See (Scheme 5).


https://doi.org/10.1016/B978-0-12-822262-1.00003-2

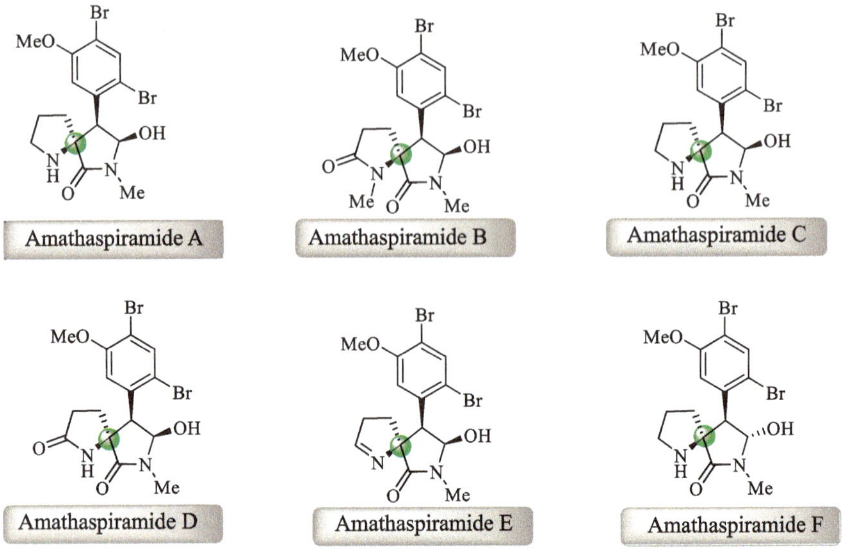


**FIGURE 3** Structures of amathaspiramides A–F.

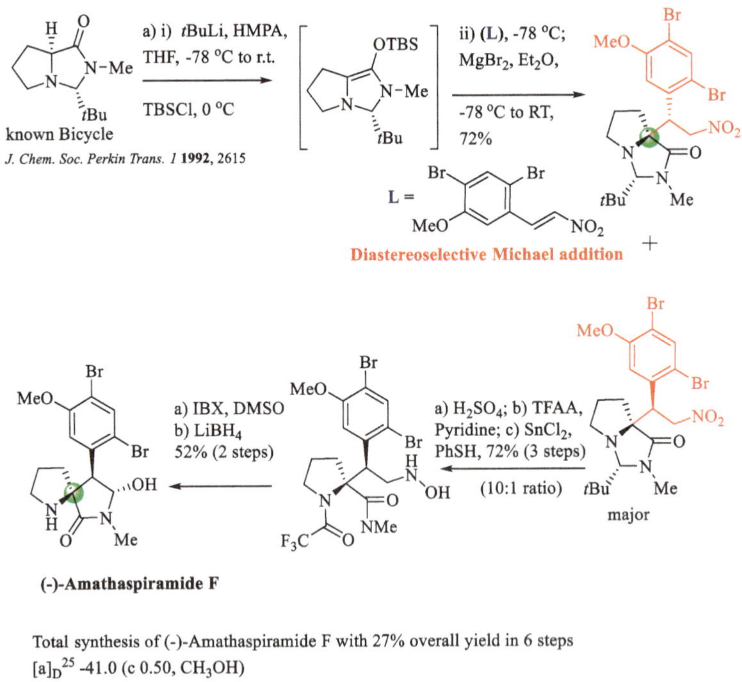


**SCHEME 5** Total synthesis of (–)-amathaspiramide F via Micheal addition.[2]

## 3.3 Ohfune's (−)-total synthesis of (−)-amathaspiramide F (2008)

See (Scheme 6).

## 3.4 Fukuyama's total syntheses of (−)-amathaspiramides (2012)

See (Scheme 7).

## 3.5 Tambar's formal synthesis of (±)-amathaspiramide F (2013)

See (Scheme 8).

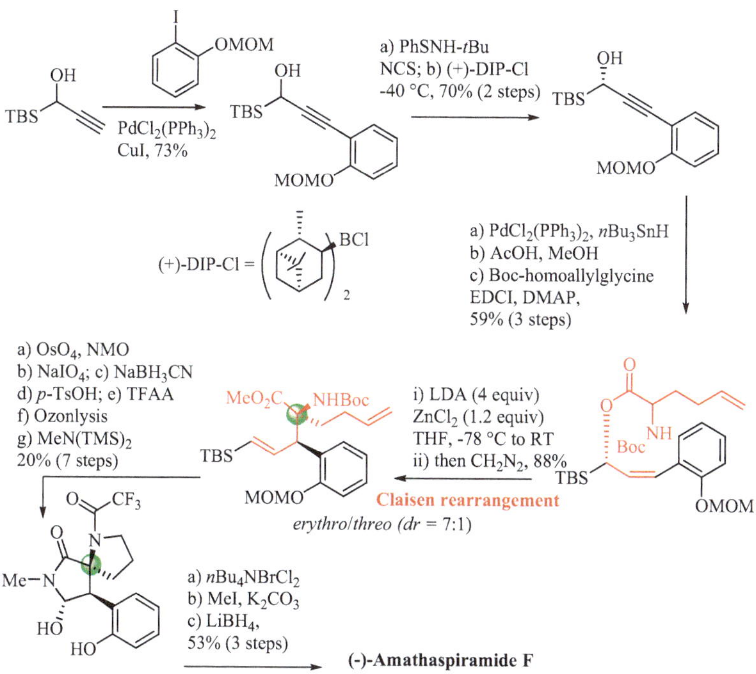


Total synthesis of (-)-Amathaspiramide F with overall yield 1.3% in 17 steps

$[\alpha]_D^{24}$ −39.0 (*c* 0.30, MeOH)

**SCHEME 6** Total synthesis via Claisen rearrangement.[3]

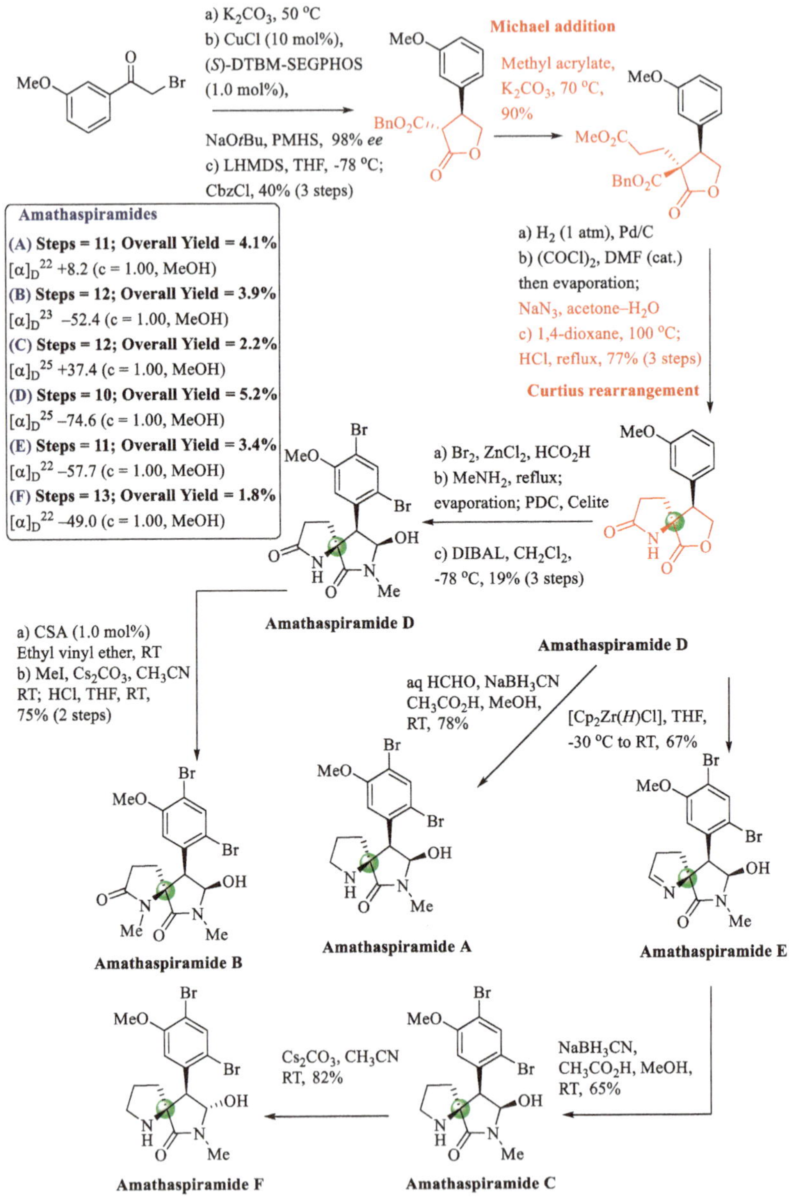


SCHEME 7 Total syntheses of (−)-amathaspiramides via Micheal addition.[4]

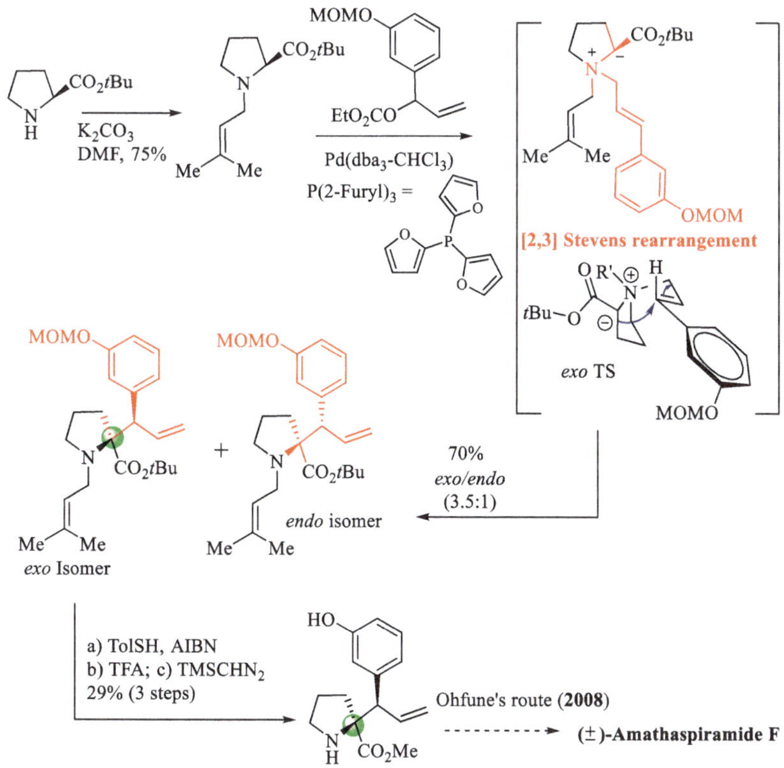


SCHEME 8 Formal synthesis of (±)-amathaspiramide F via [2,3]-Stevens rearrangement.[5]

## 3.6 Lee's synthesis of amathaspiramide C (2015)

See (Scheme 9).

## 3.7 Sun's synthesis of amathaspiramides B, D, and F (2016)

See (Scheme 10).

## 3.8 Kim's synthesis of (−)-amathaspiramide F (2018)

See (Scheme 11).

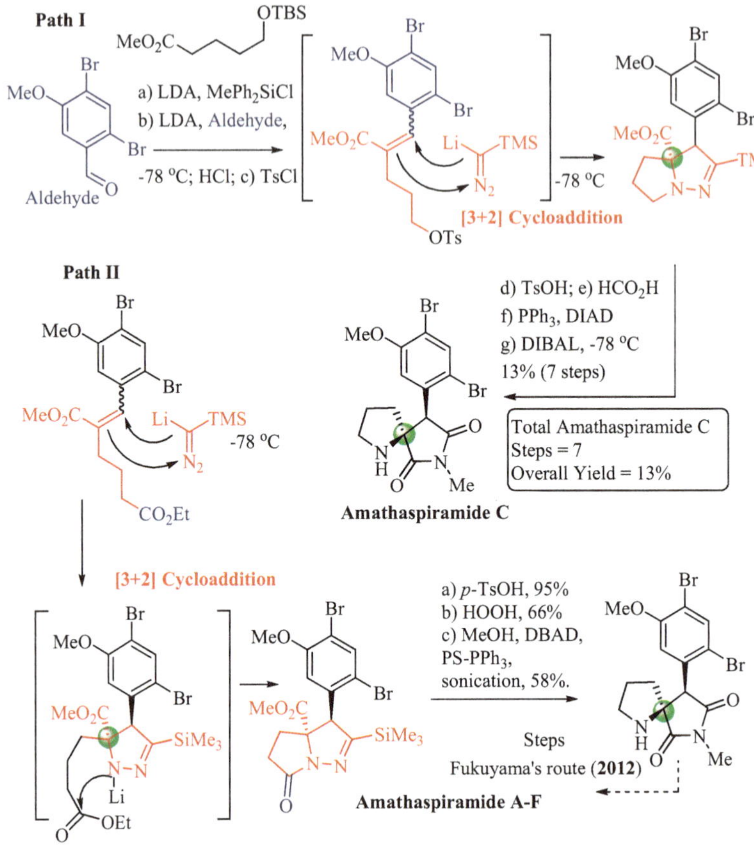


SCHEME 9 Amathaspiramide C via [3 + 2]-cycloaddition.[6]

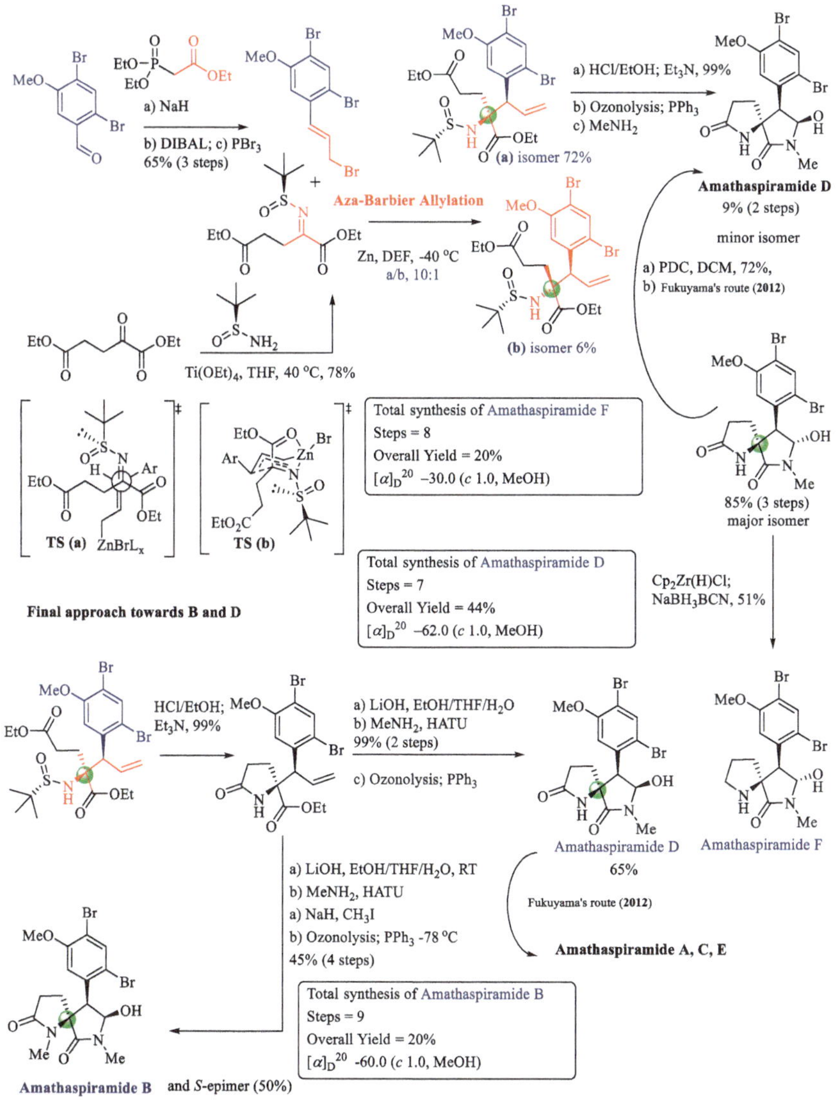


**SCHEME 10** Amathaspiramides B, D, and F via aza-Barbier Allylation.

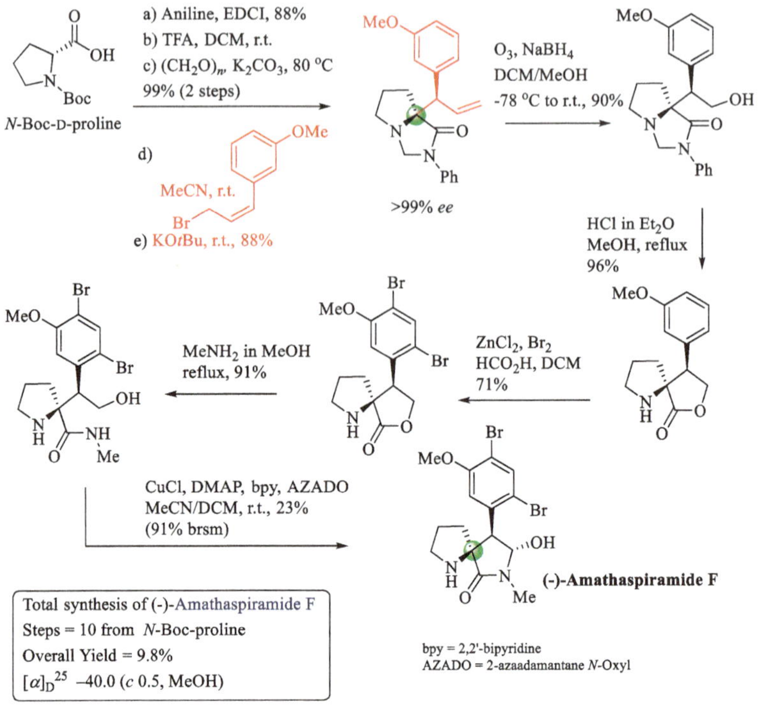


**SCHEME 11** (–)-Amathaspiramide F via asymmetric Cα-alkylation of proline.[7]

# References


1. Morris, B. D.; Prinsep, M. R.; Amathaspiramides, A.–F. Novel Brominated Alkaloids from the Marine Bryozoan Amathia Wilsoni. *J. Nat. Prod.* **1999,** *62,* 688–693.
2. Hughes, C. C.; Trauner, D. The Total Synthesis of (–)-Amathaspiramide F. *Angew. Chem. Int. Ed.* **2002,** *41,* 4556–4559. and references therein.
   i) Moss, W. O.; Jones, A. C.; Wisedale, R.; Mahon, M. F.; Molloy, K. C.; Bradbury, R. H.; Hales, N. J.; Gallagher, T. 2-Amino Ketene S, S-Acetals as α-amino Acid Homoenolate Equivalents. Synthesis of 3-substituted Prolines and Molecular Structure of 2-(*n*-Pivaloyl-pyrrolidin-2-Ylidene)-1, 3-dithiane. *J. Chem. Soc. Perkin Trans.* **1992,** *1,* 2615–2624.
3. Sakaguchi, K.; Ayabe, M.; Watanabe, Y.; Okada, T.; Kawamura, K.; Shinada, T.; Ohfune, Y. Total Synthesis of (–)-amathaspiramide F, *Tetrahedron* **2009,** *65,* 10355–10364.
   Sakaguchi, K.; Ayabe, M.; Watanabe, Y.; Okada, T.; Kawamura, K.; Shiada, T.; Ohfune, Y. Total synthesis of (–)-amathaspiramide F, Org. Lett. 2008, 10, 5449-5452.
4. Chiyoda, K.; Shimokawa, J.; Fukuyama, T. Total Syntheses of All the Amathaspiramides. *Angew. Chem. Int. Ed.* **2012,** *124,* 2555–2558.
5. Soheili, A.; Tambar, U. K. Synthesis of (±)-amathaspiramide F and Discovery of an Unusual Stereocontrolling Element for the [2, 3]-Stevens Rearrangement. *Org. Lett.* **2013,** *15,* 5138–5141.
6. O'Connor, M.; Sun, C.; Lee, D. Synthesis of Amathaspiramides by Aminocyanation of Enoates. *Angew. Chem. Int. Ed.* **2015,** *54,* 9963–9966.
7. Cho, H.; Shin, J. E.; Lee, S.; Jeon, H.; Park, S.; Kim, S. Asymmetric Cα-Alkylation of Proline *via* Chirality Transfers of Conformationally Restricted Proline Derivative: Application to the Total Synthesis of (–)-Amathaspiramide. *F.* Org. Lett. **2018,** *20,* 6121–6125.

CHAPTER

# 4

# Cephalotaxine

## 4.1 Abstract

Cephalotaxine (CET) is a parent polycyclic scaffold for the class of *Cephalotaxus alkaloids*.[1] The ester derivatives of CET, i.e., harringtonine (**2**), isoharringtonine (**3**), homoharringtonine (**4**), and deoxyharringtonine (**5**) show important anticancer activity against different types of leukemia (Fig. 4).[2] In 1963, Paudler and his coworker[3] isolated it from *Cephalotaxus drupacea* and *Cephalotaxus fortunei*. Later on, the stereochemistry of CET was established by Powell et al.[4,5] via X-ray crystallography. The natural product (CET) possesses unique structural framework that consists of pentacyclic ring system with three contiguous stereogenic centers. The presence of quaternary center next to nitrogen is one of the key synthetic challenges in the synthesis of this CET. The interesting structural features of natural product and prominent biological activities of its ester derivatives make it attractive synthetic target for chemists to develop novel synthetic strategies and efficient synthetic routes.

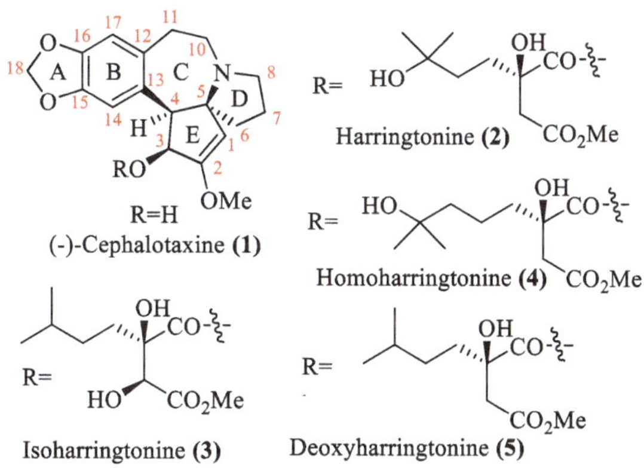


FIGURE 4 Structures of cephalotaxine and its ester derivatives.


https://doi.org/10.1016/B978-0-12-822262-1.00004-4

## 4.2 Biosynthesis

See (Scheme 12).

## 4.3 Weinreb's first total (±)-cephalotaxine synthesis (1975)

See (Scheme 13).

- Lewis acid [$BF_3 \bullet (Et_2O)_2$] catalyzes cyclization to closed seven-membered benzazepine ring.
- Magnesium catalyzed intramolecular Michael reaction to furnish pentacycle (±)-demethylcephalotaxine.

## 4.4 Semmelhack's total synthesis of cephalotaxine (1975)

See (Scheme 14).

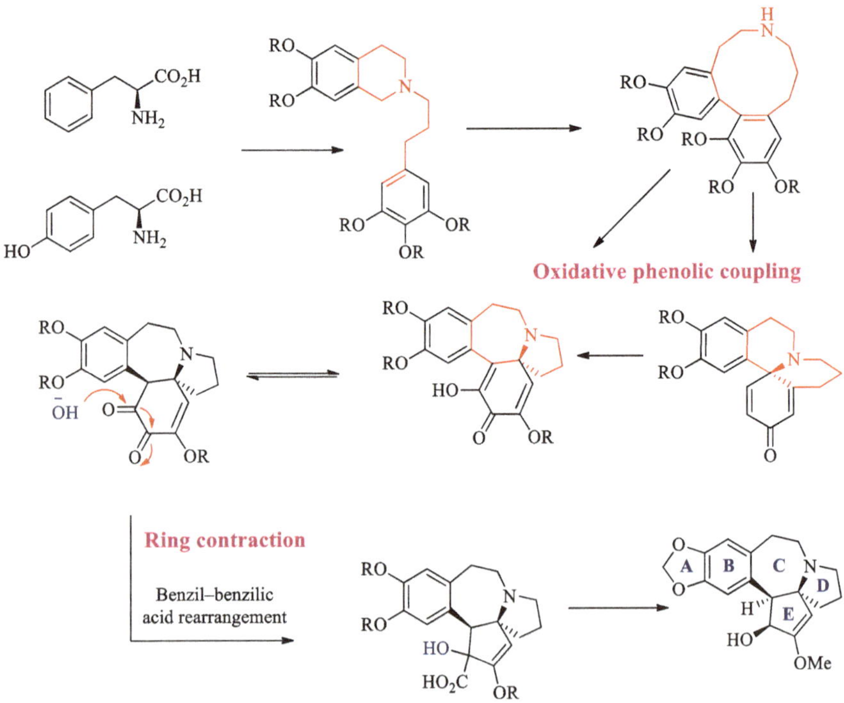


SCHEME 12 Biosynthesis of cephalotaxine.[6]

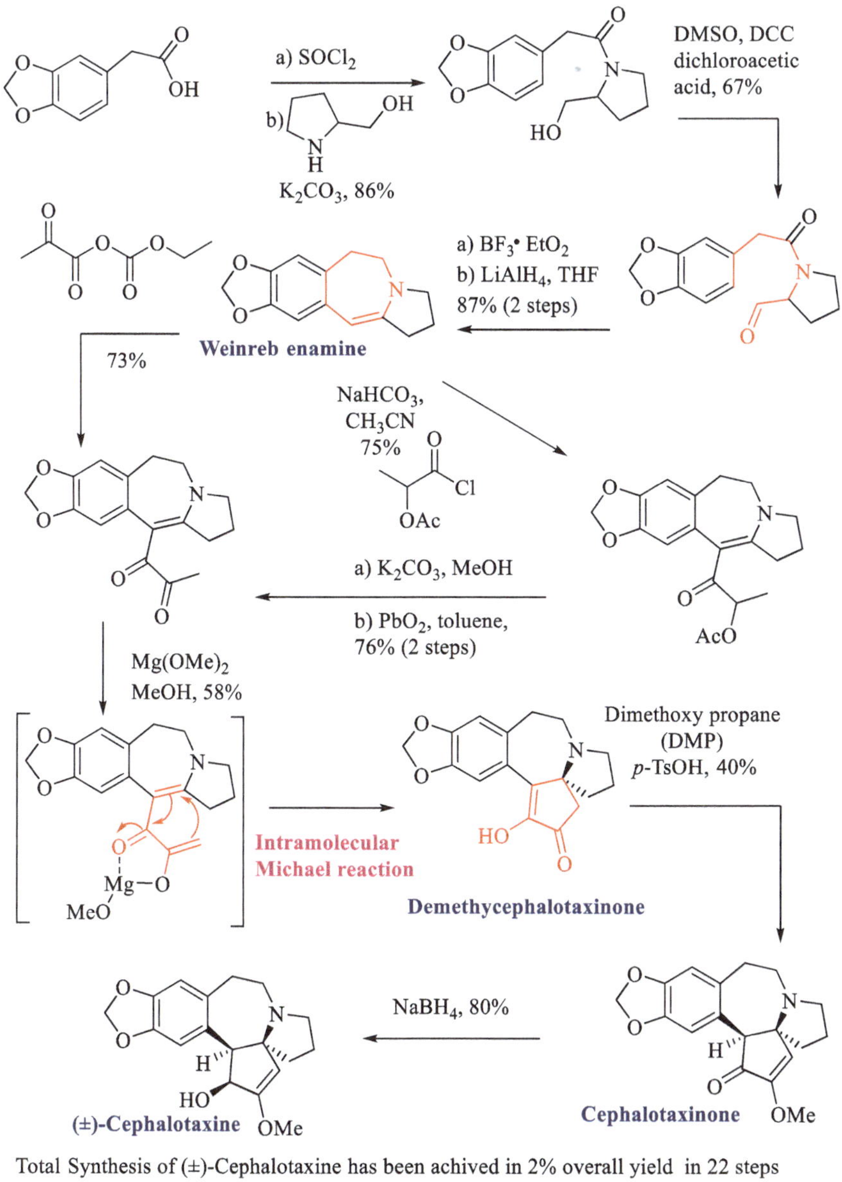


SCHEME 13 First total (±)-cephalotaxine synthesis.[7]

## 4.5 Hanaoka's first-generation (±)-total synthesis (1986)

See (Scheme 15).

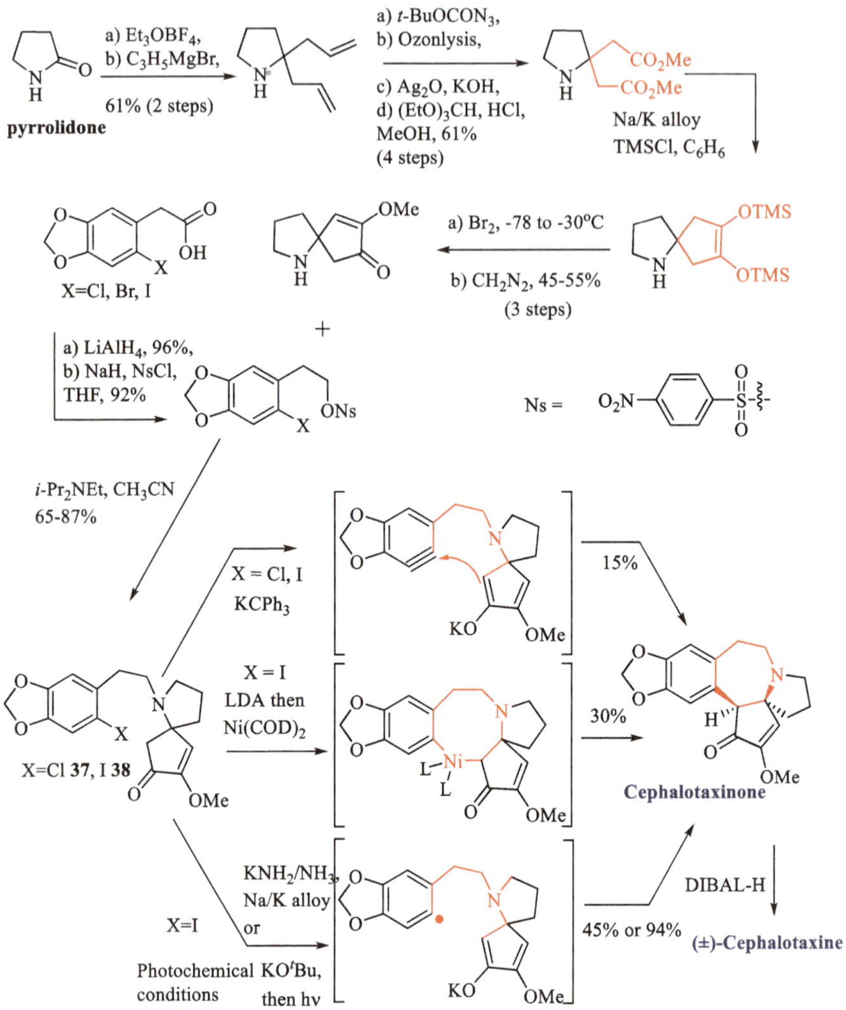


Total Synthesis (±)-Cephalotaxine in 12-13% overall yield in 11 steps from pyrrolidone

SCHEME 14 Total synthesis of cephalotaxine, a convergent strategy.[8]

## 4.6 Hanaoka's second-generation formal synthesis (1988)

See (Scheme 16).

## 4.7 Kuehne's total synthesis (1988)

See (Scheme 17).

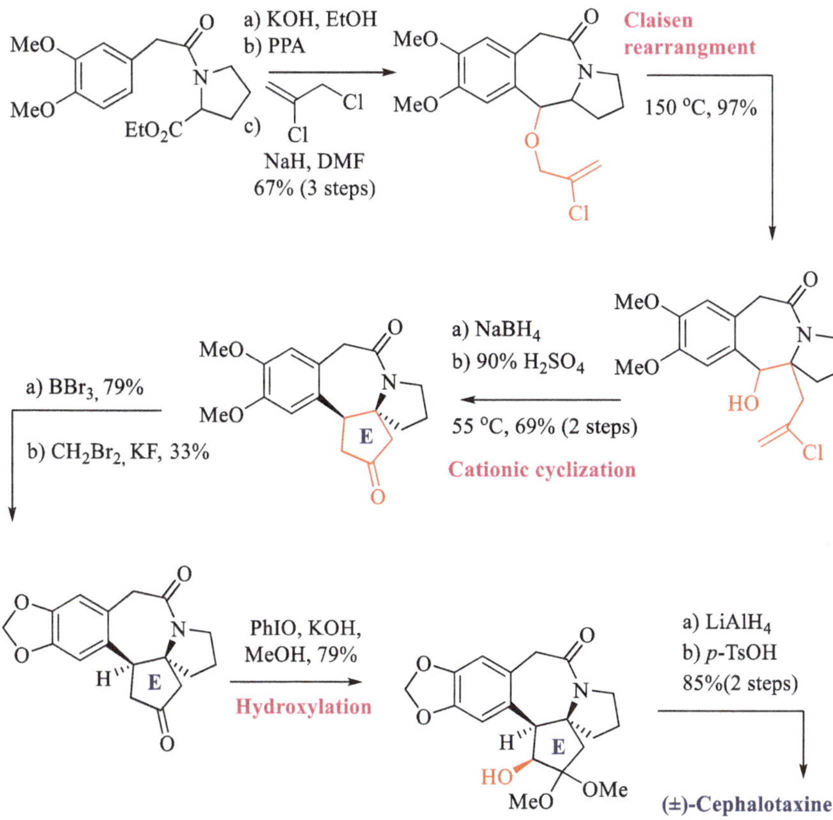


(±)-Cephalotaxine synthesis in approximately 4.5% overall yield in 10 steps

SCHEME 15 Cephalotaxine (±)-total synthesis via Claisen rearrangement.[9]

## 4.8 Fuchs's total synthesis of (±)-cephalotaxine (1988)

See (Scheme 18).

## 4.9 Ikeda's total racemic synthesis (1990/1993)

See (Scheme 19).

## 4.10 Mori's asymmetric (−)-cephalotaxine synthesis (1995)

See (Scheme 20).

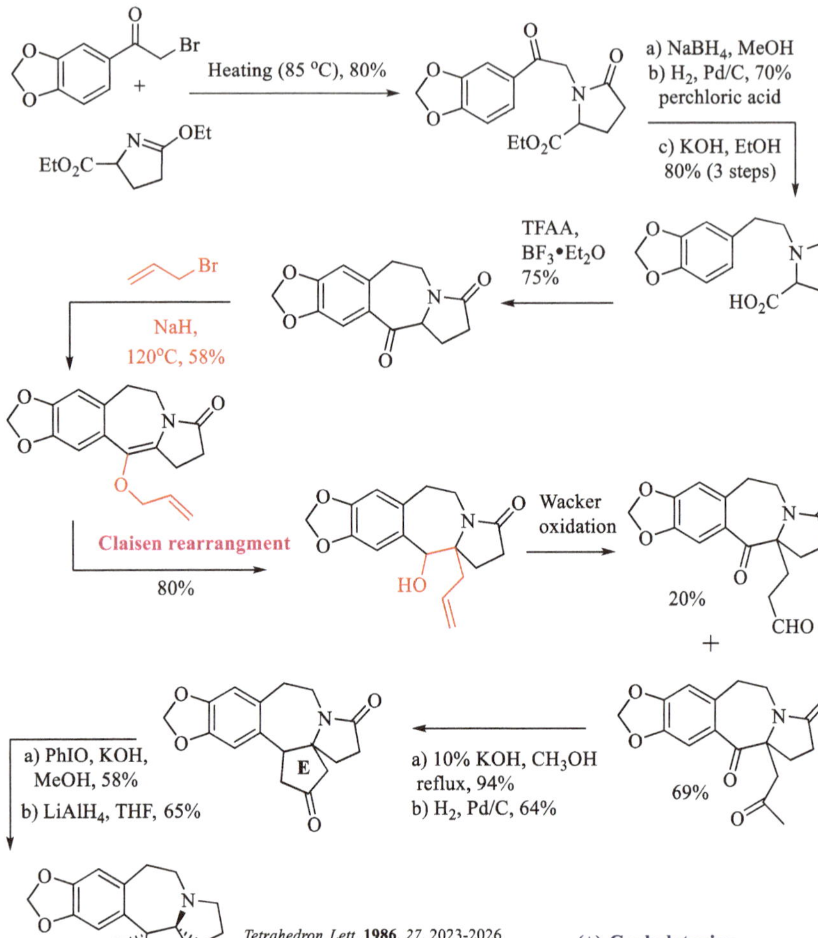


SCHEME 16 Second-generation formal synthesis of cephalotaxine.[10]

## 4.11 Mariano's synthesis *via* two interrelated strategies (1996)

**(I)** Single-electron transfer (SET)—promoted photocyclization

See (Scheme 21).

**(II)** Transannular cyclization approach

See (Scheme 22).

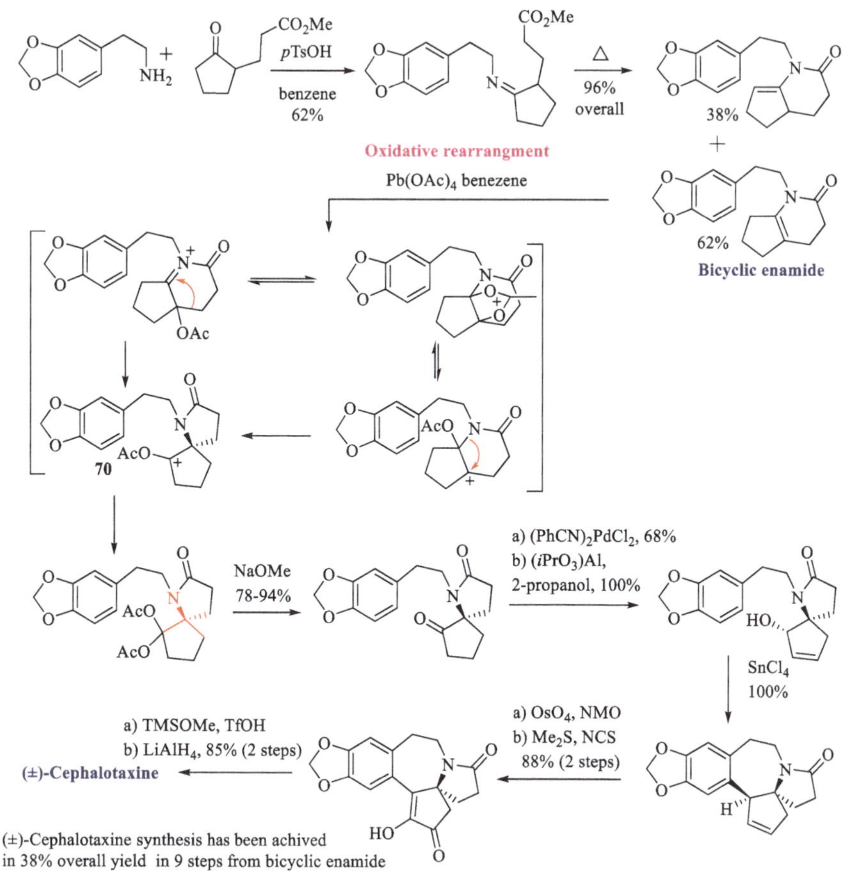


SCHEME 17 Total synthesis via novel oxidative rearrangement.[11]

## 4.12 Nagasaka's synthesis of (−)-cephalotaxine (1997)

See (Scheme 23).

## 4.13 El Bialy's formal synthesis (1998)

See (Scheme 24).

## 4.14 Ikeda's formal synthesis (1999)

See (Scheme 25).

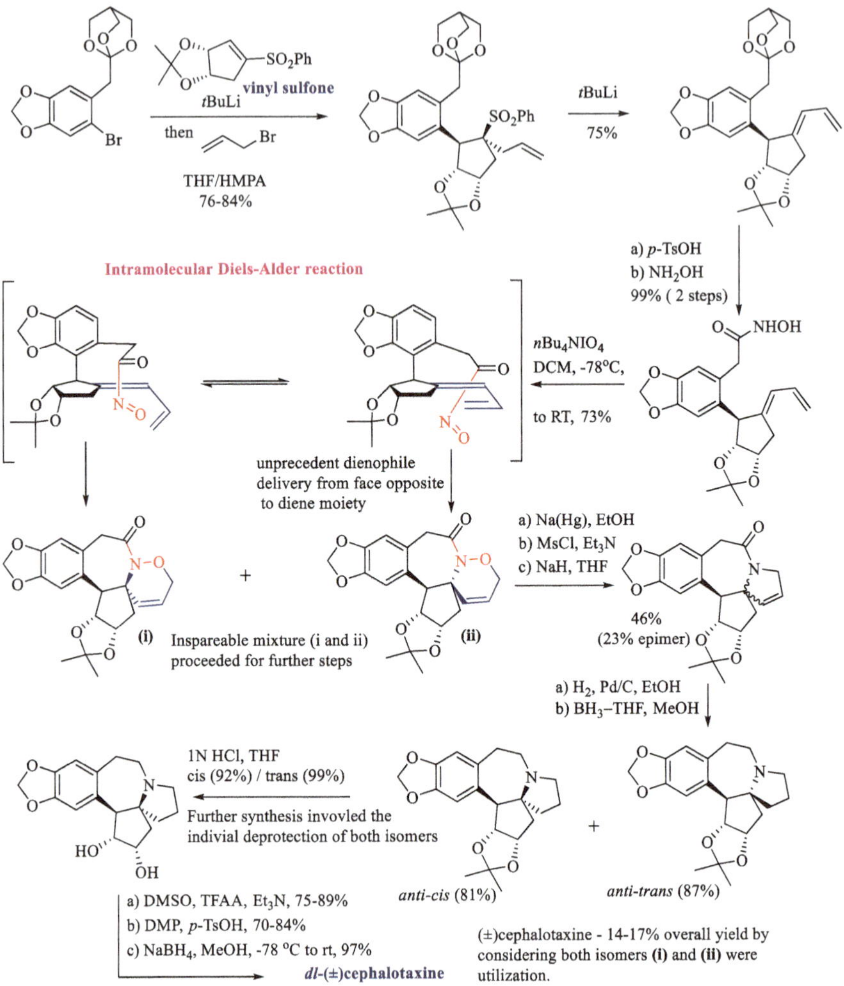


SCHEME 18 Total synthesis via intramolecular Hetero Diels–Alder reaction.[12]

## 4.15 Tietze's synthetic approach (1999)

See (Scheme 26).

- Heck reaction by using Hermanto's catalyst[21] to form the benzazepine ring and to complete pentacycle.

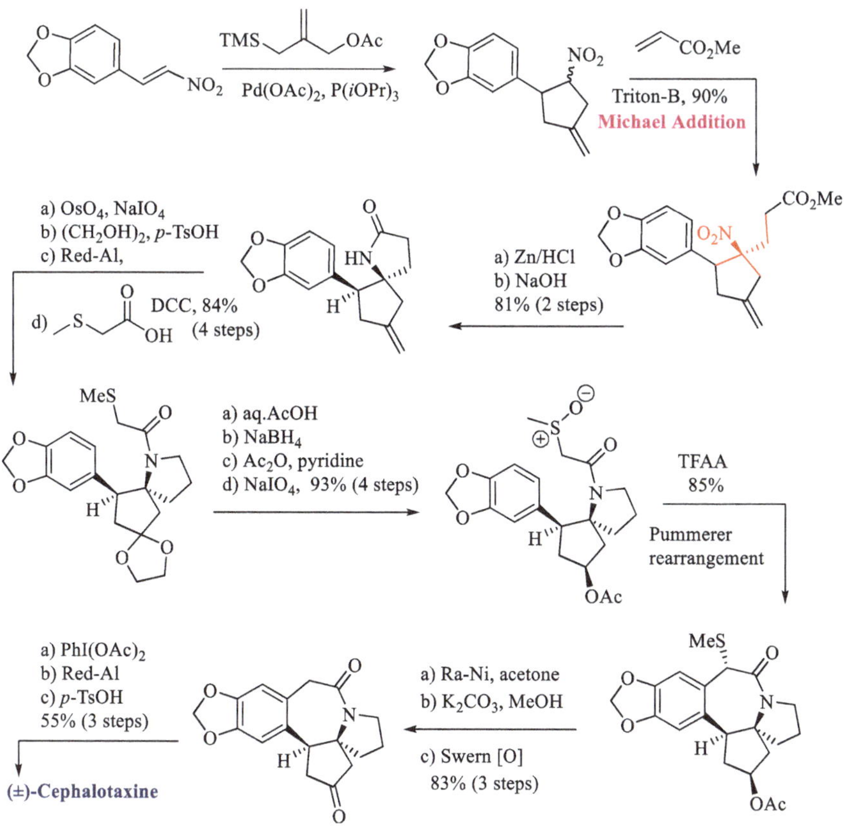


SCHEME 19 Synthesis via acid-catalyzed Pummerer rearrangement.[13]

## 4.16 Nagasaka's synthesis (2002)

See (Scheme 27).

## 4.17 Yoshida's formal synthesis (2002)

See (Scheme 28).

## 4.18 Li's synthesis (2003)

**(I)** *Via* transannular reductive rearrangement

See (Scheme 29).

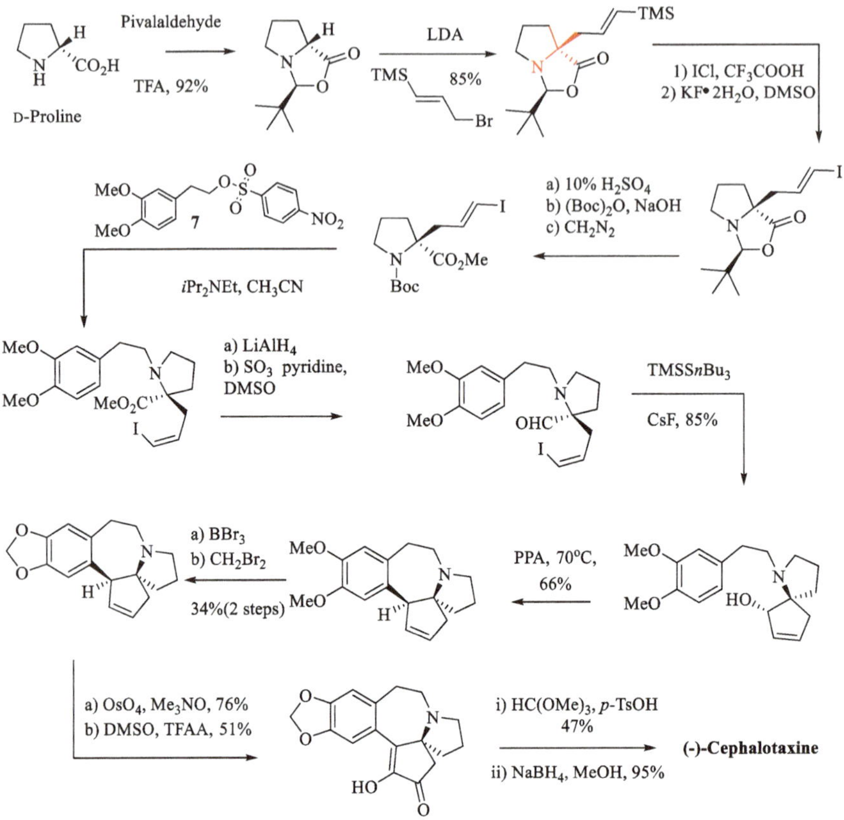


(-)-Cephalotaxine asymmetric total synthesis was completed in in 19 steps with approximately 1% overall yield

SCHEME 20 First asymmetric synthesis of (−)-cephalotaxine via alkylation by employing Seebach-type[14] chemistry.[15]

**(II)** *Via unusual azo-Nazarov-type cyclization*

See (Scheme 30).

## 4.19 Royer's synthesis *via* semipinacolic rearrangement (2004)

See (Scheme 31).

## 4.20 Li's second-generation formal synthesis (2005)

See (Scheme 32).

***I) Single-Electron-Transfer (SET) promoted photocyclisation***

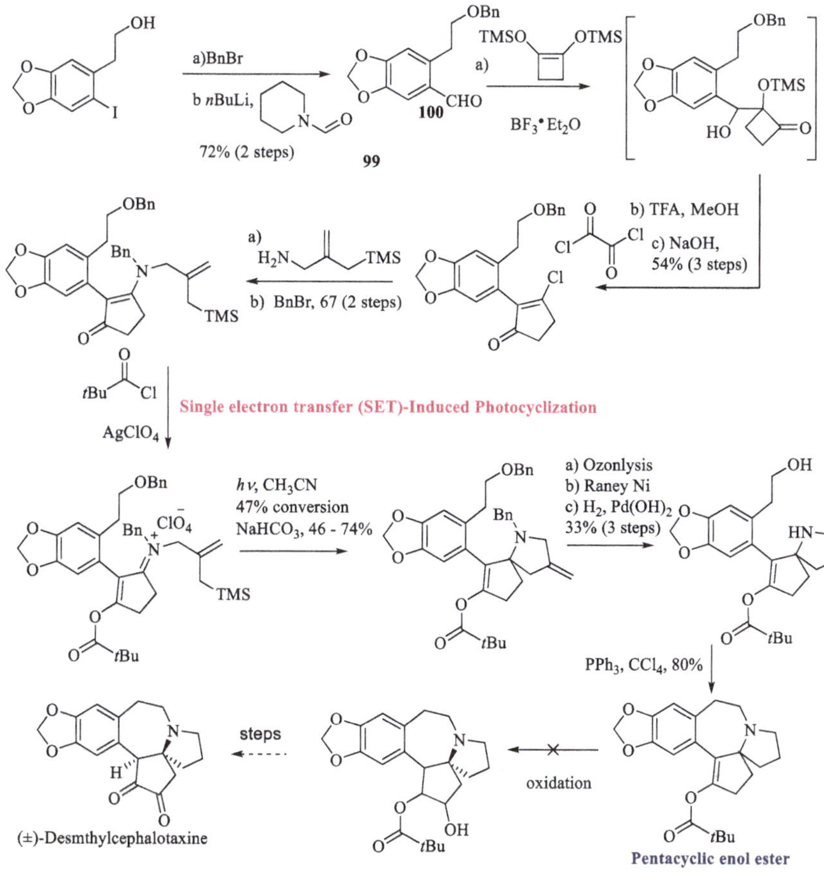


SCHEME 21 Synthetic route toward cephalotaxine synthesis.

## 4.21 Li's synthesis of Dolby–Weinreb enamine

See (Scheme 33).

## 4.22 Mariano's formal synthesis via photocyclization reaction (2006)

See (Scheme 34).
See (Scheme 35).

***II) Transannular cyclisation Approach***

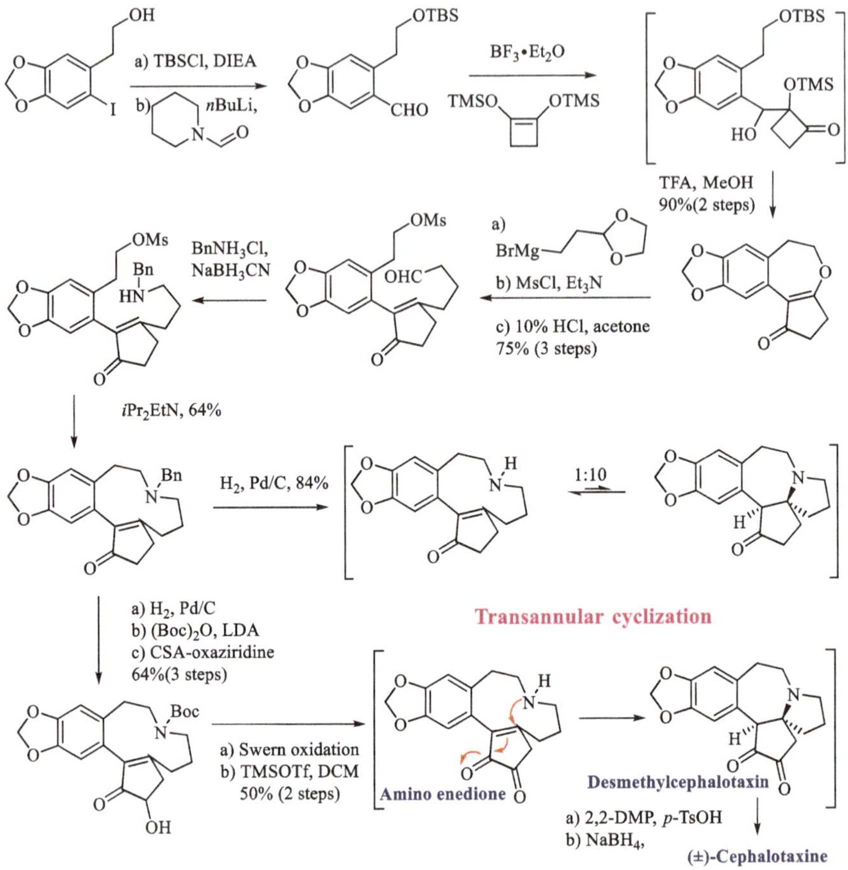


SCHEME 22 Two interrelated strategies toward cephalotaxine synthesis.[16,17]

## 4.23 Gin's synthetic studies (2006)

See (Scheme 36).

## 4.24 Li's formal synthesis (2007)

See (Scheme 37).

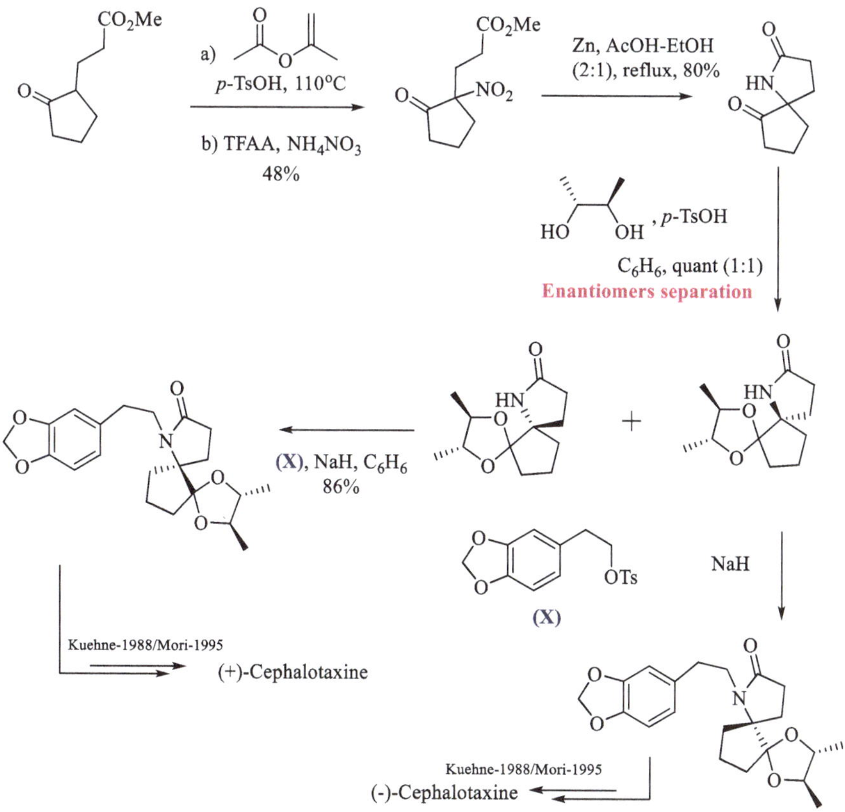


SCHEME 23 (−)-Cephalotaxine synthesis via enantiomers separation.[18]

## 4.25 Stoltz's formal synthesis via Pd-catalyzed aerobic oxidative heterocyclization chemistry (2007)

See (Scheme 38).

## 4.26 Ishibashi's total synthesis (2008)

See (Scheme 39).

## 4.27 Hayes's first formal synthesis (2008)

See (Scheme 40).

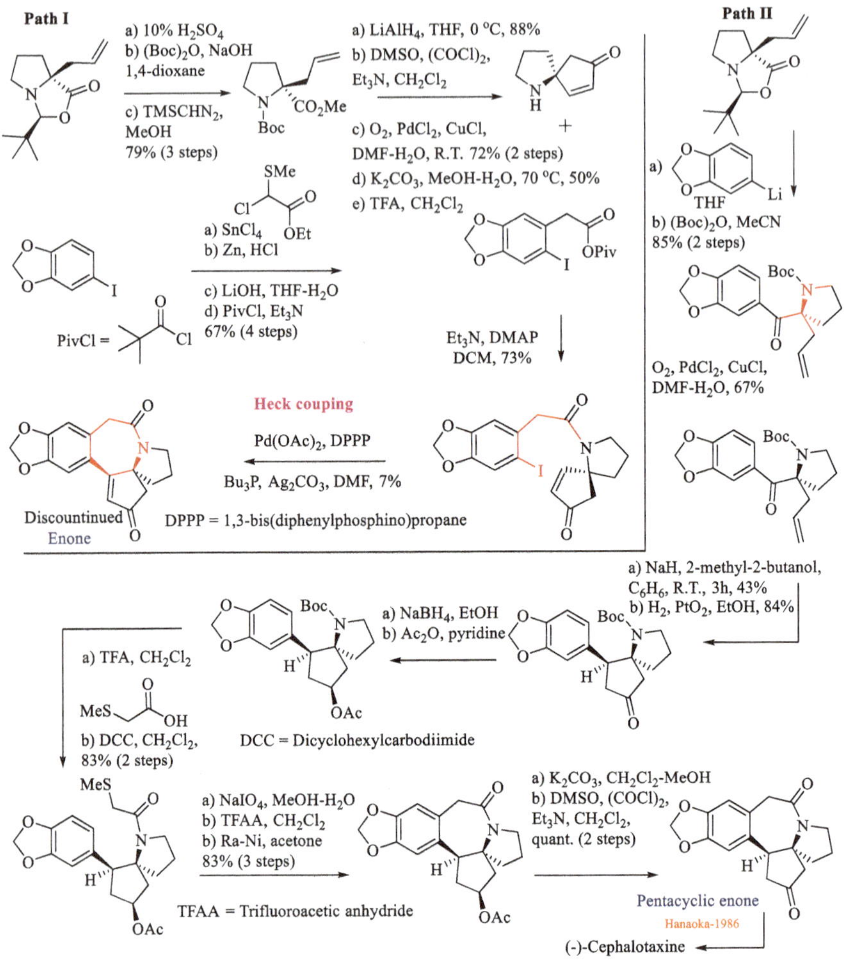


**SCHEME 24** Formal total synthesis of (−)-cephalotaxine.[19]

## 4.28 Hayes's second formal synthesis via 1,5-CH insertion reaction (2008)

See (Scheme 41).

## 4.29 Bubnov's approach toward cephalotaxine (2008)[33]

See (Scheme 42).

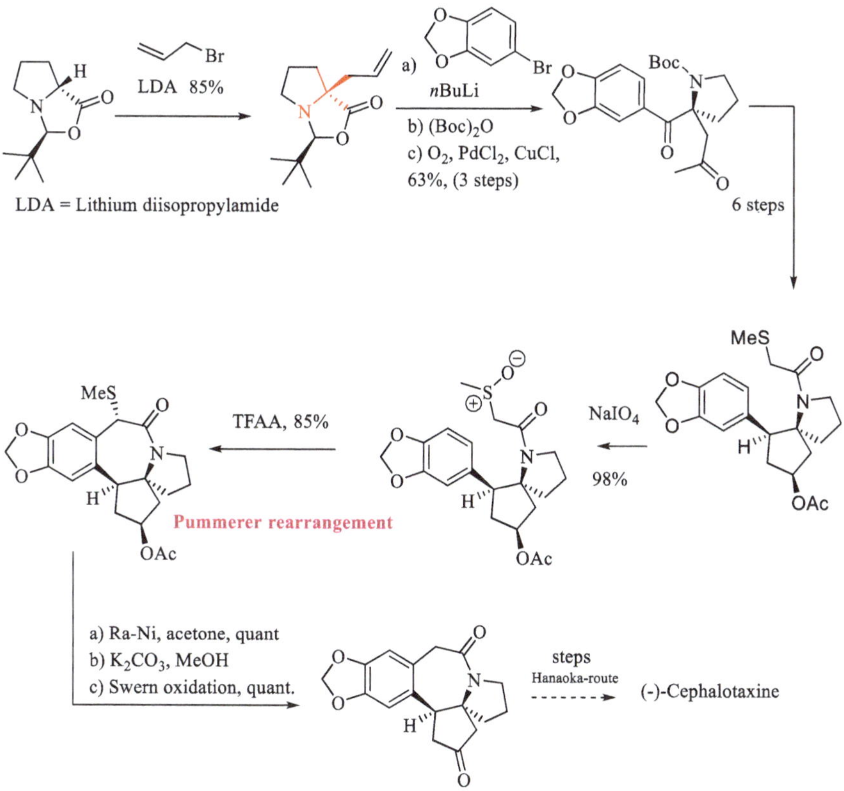


SCHEME 25 Cephalotaxine formal synthesis via Pummerer rearrangement reaction.[19]

## 4.30 Liu's formal synthesis (2009)

See (Scheme 43).
See (Scheme 44).

## 4.31 Zhang synthesis (2009)

See (Scheme 45).

## 4.32 Li's total synthesis (2011)

See (Scheme 46).

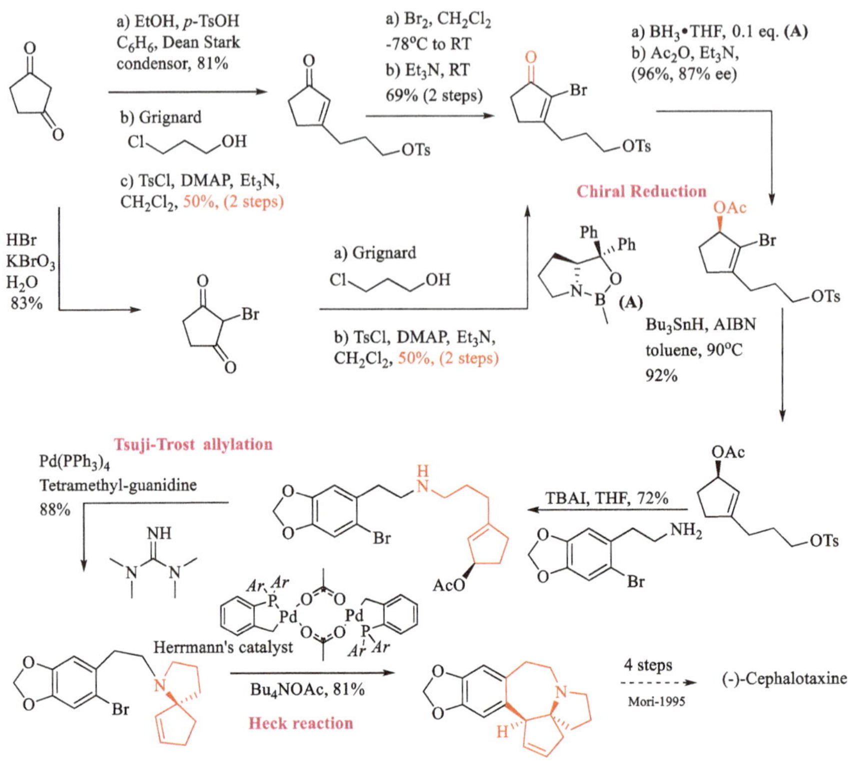


SCHEME 26 Cephalotaxine' synthesis via palladium-catalyzed transformations.[20]

## 4.33 Tu's (−)-formal synthesis (2012)

See (Scheme 47).

## 4.34 Renaud's (−)-formal synthesis (2012)

See (Scheme 48).

## 4.35 Zhang-Liu's formal synthesis (2012)

See (Scheme 49).
See (Scheme 50).

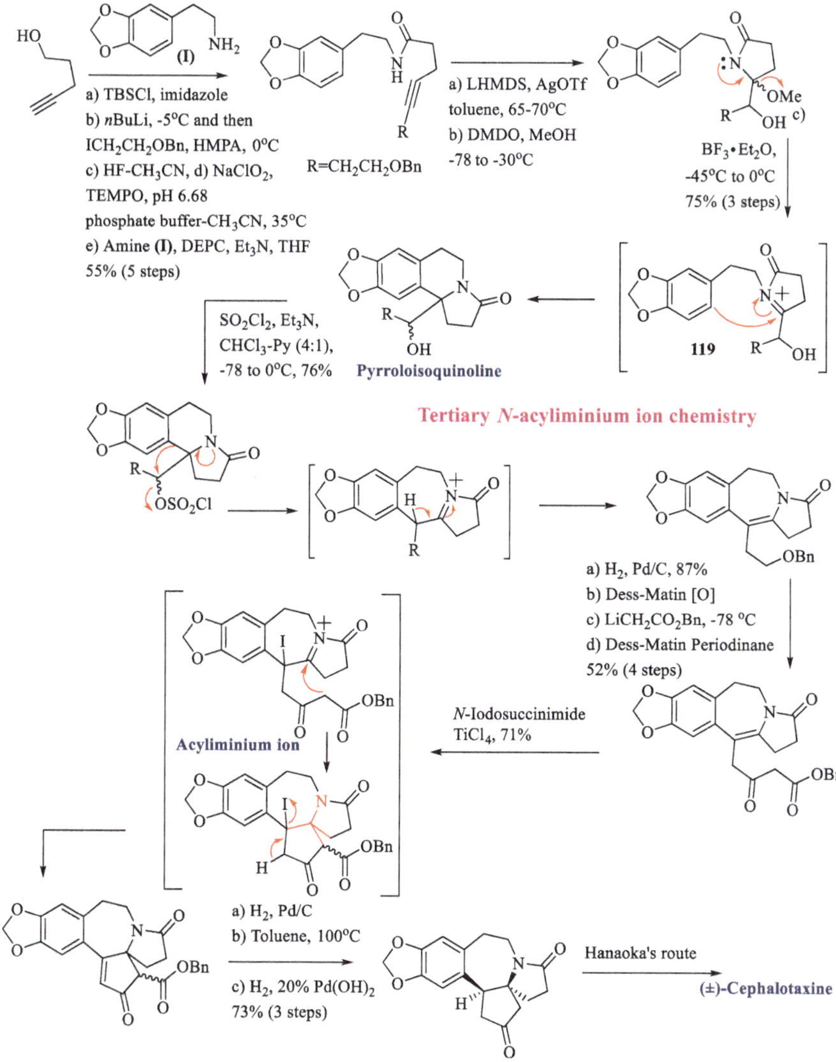


**SCHEME 27** Cephalotaxine synthesis via tertiary N-acyliminium ion chemistry.[18]

## 4.36 Jiang's formal synthesis (2013)

See (Scheme 51).

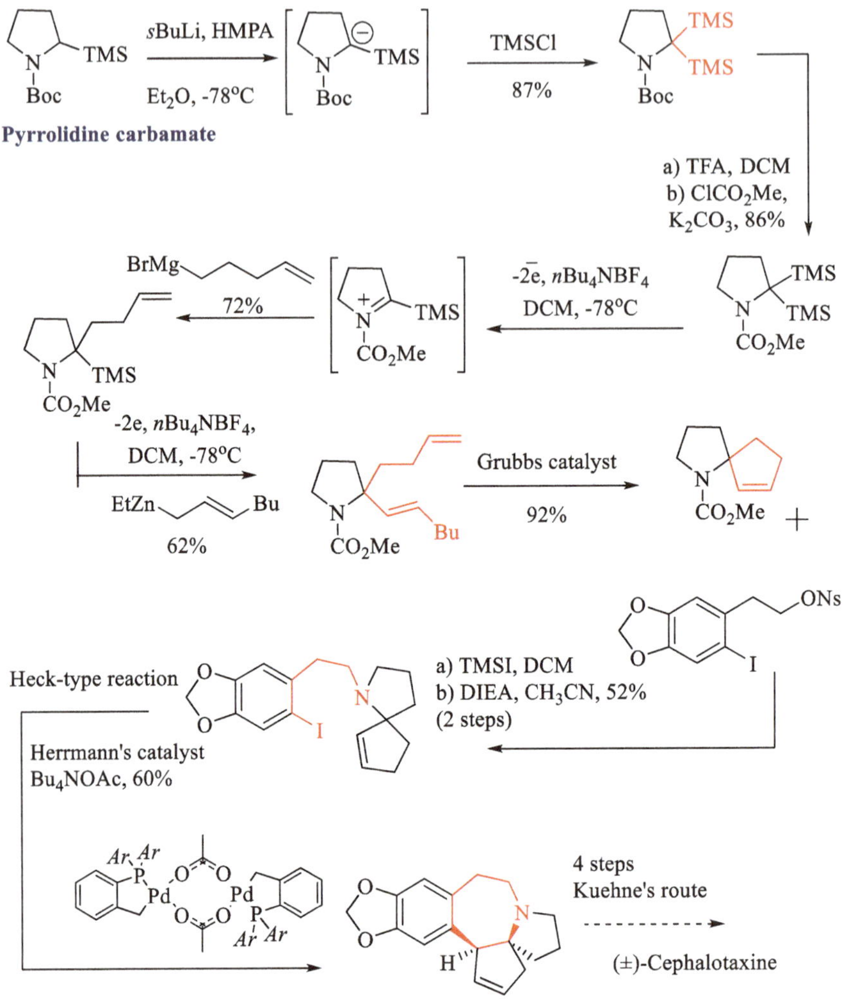


SCHEME 28 Cephalotaxine formal synthesis.[22]

## 4.37 Huang's formal synthesis (2013)

See (Scheme 52).

## 4.38 Huang's formal synthesis (2015)

See (Scheme 53).

### *I) Via* transannular reductive rearrangement

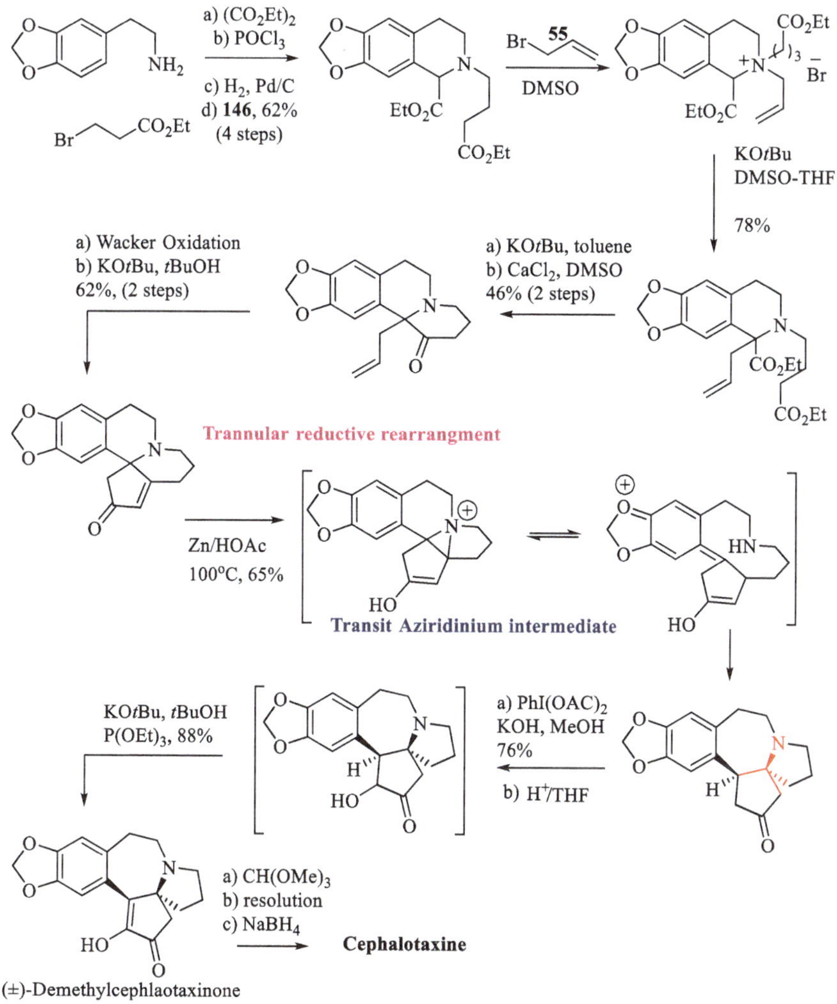


SCHEME 29 Cephalotaxine synthesis.[23]

## 4.39 Hong's formal synthesis (2015)

See (Scheme 54).

***II) Via unusual azo-Nazarov-type cyclization***

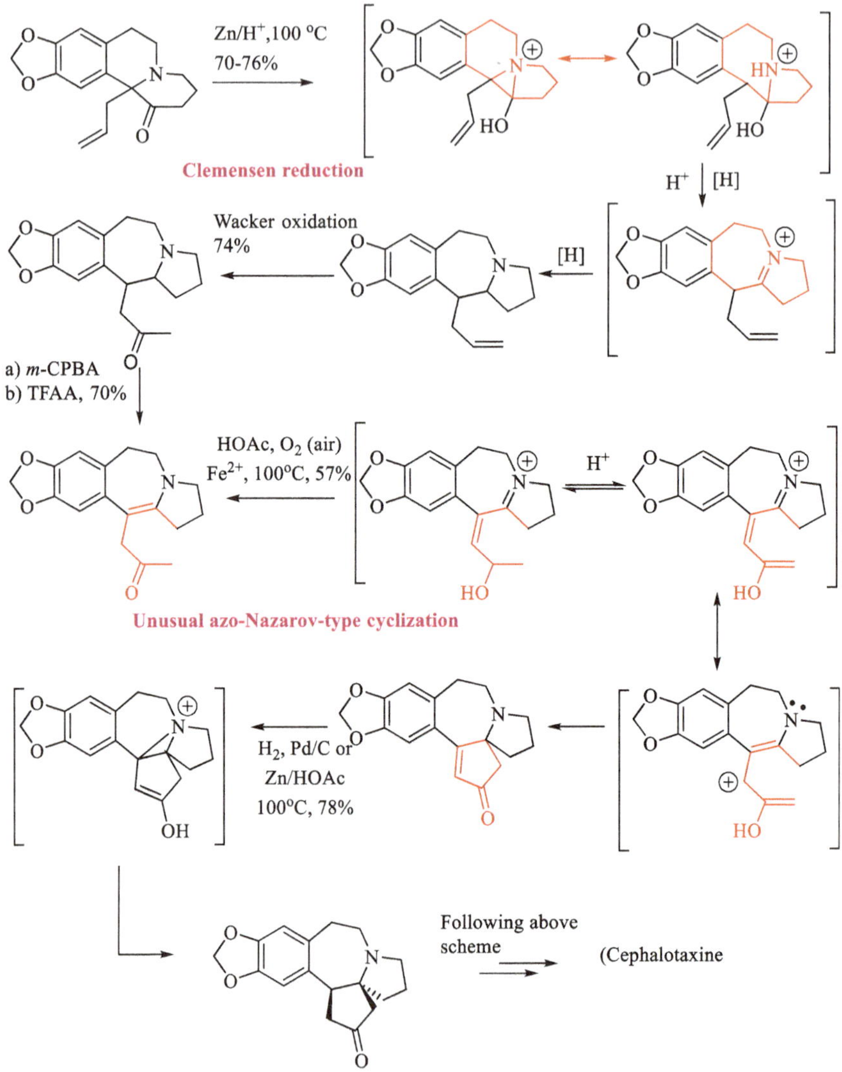


SCHEME 30 Cephalotaxine via unusual azo-Nazarov-type cyclization.

## 4.40 Chandrasekhar's formal total synthesis (2016)

See (Scheme 55).

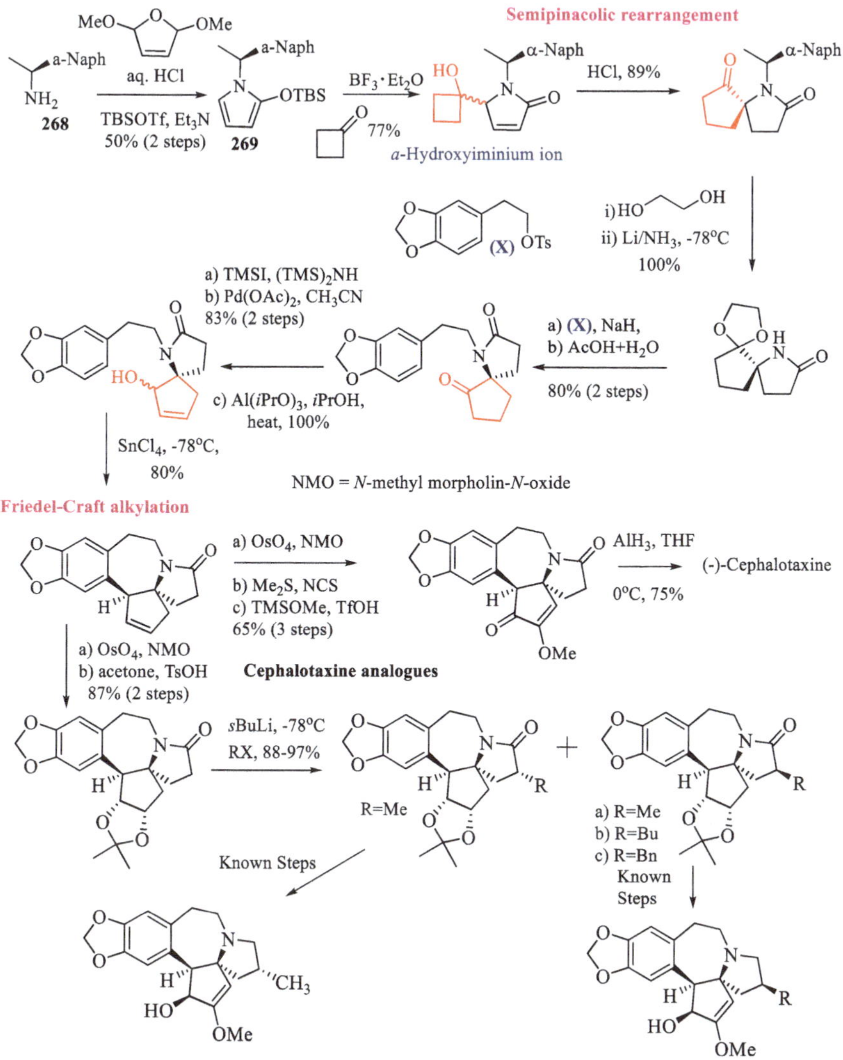


SCHEME 31 Cephalotaxine synthesis via semipinacolic rearrangement.[24]

## 4.41 Fan's total synthesis (2017)

See (Scheme 56).

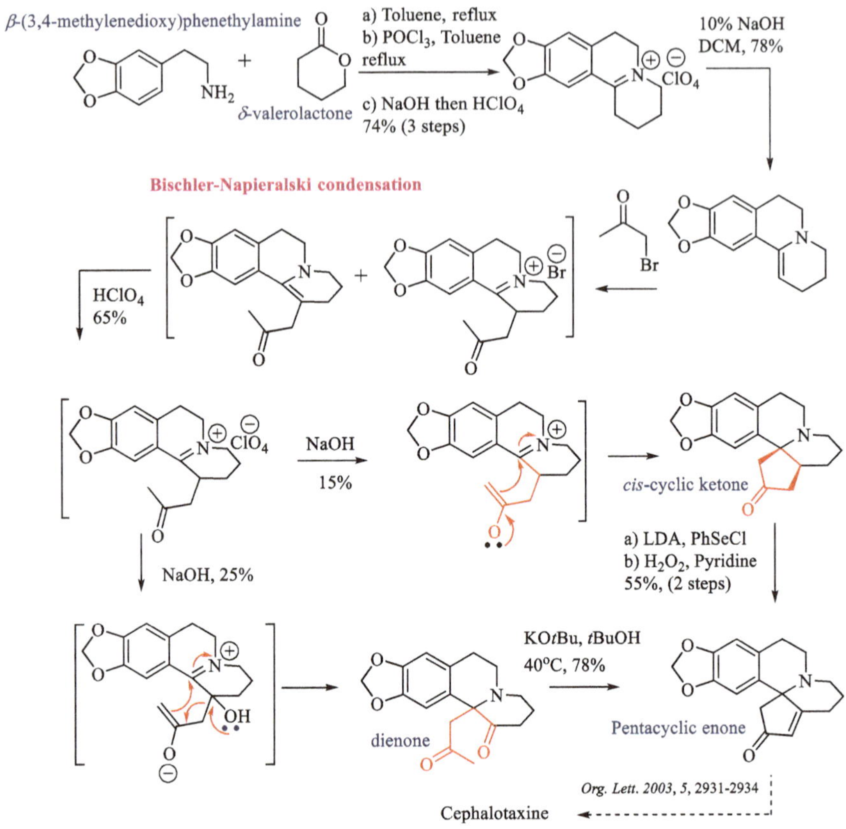


**SCHEME 32** Cephalotaxine formal synthesis.[25]

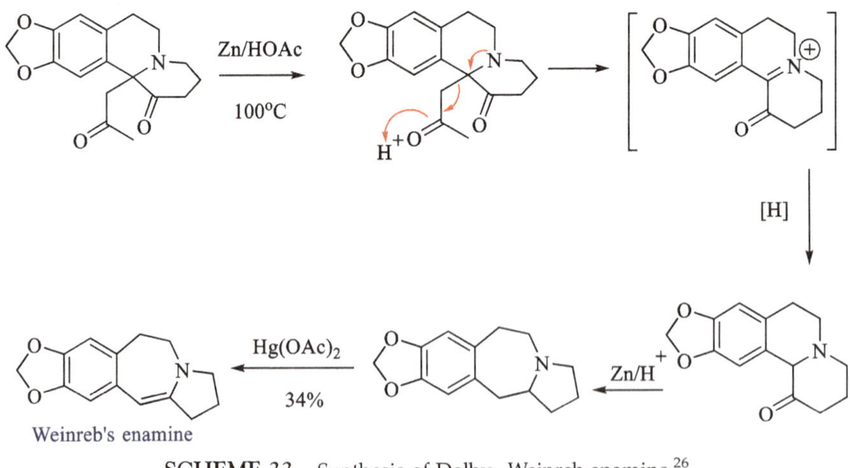


**SCHEME 33** Synthesis of Dolby–Weinreb enamine.[26]

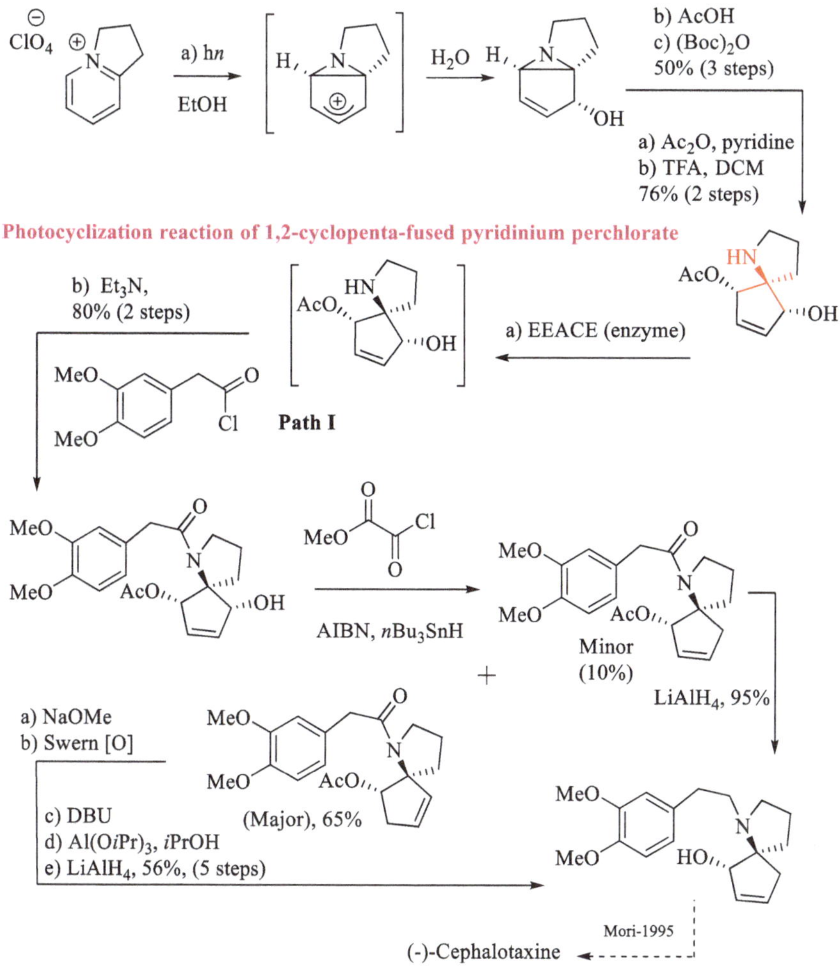


SCHEME 34 Cephalotaxine formal synthesis via photocyclization reaction.[27]

## 4.42 Beaudry's (−)-total synthesis via furan oxidation−transannular Mannich cyclization (2019)

See (Scheme 57).

## 4.43 Kim's formal (−)-total synthesis (2019)

See (Scheme 58).

**Path II**

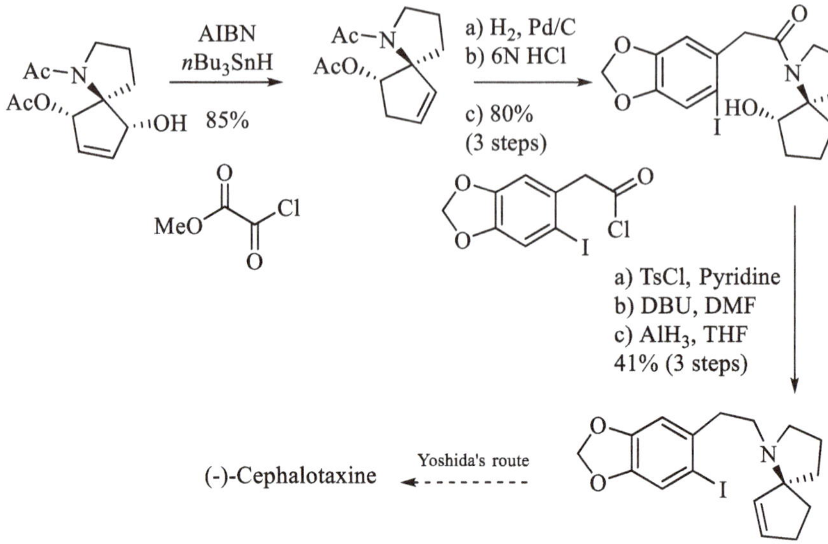


SCHEME 35 Synthetic pathway II.

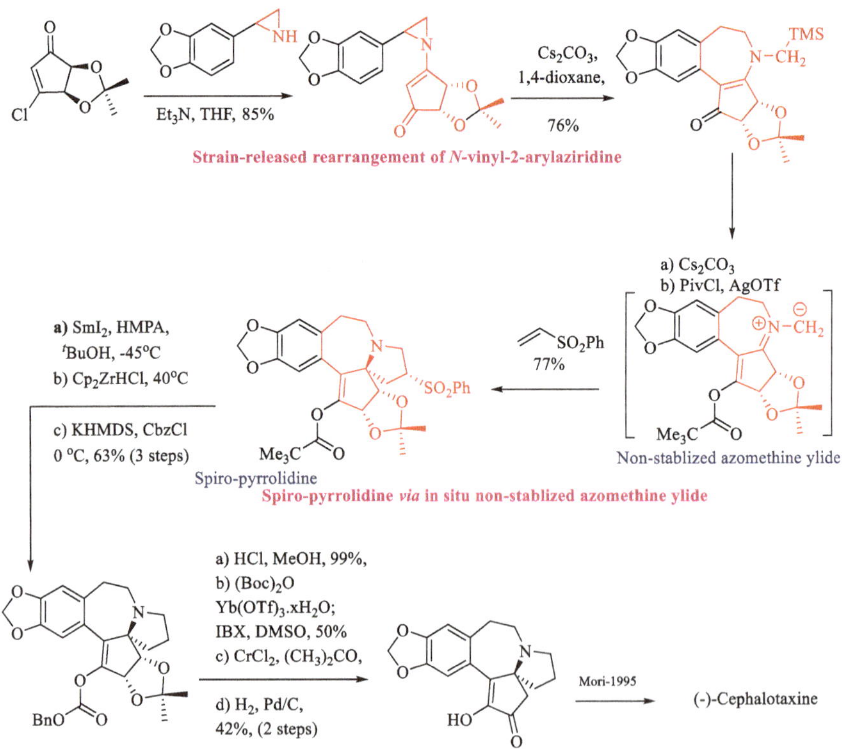


SCHEME 36 Synthesis via strain-released rearrangement of *N*-vinyl-2-arylaziridines.[28]

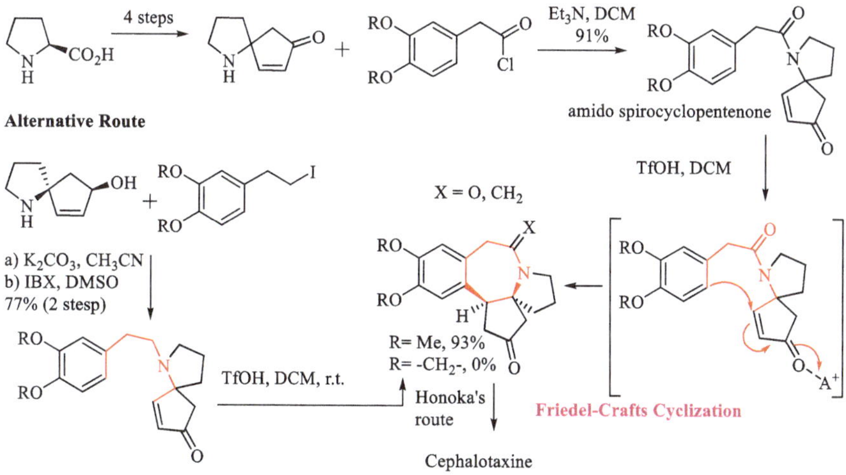


SCHEME 37 Formal synthesis via Friedel−Crafts cyclization.[29]

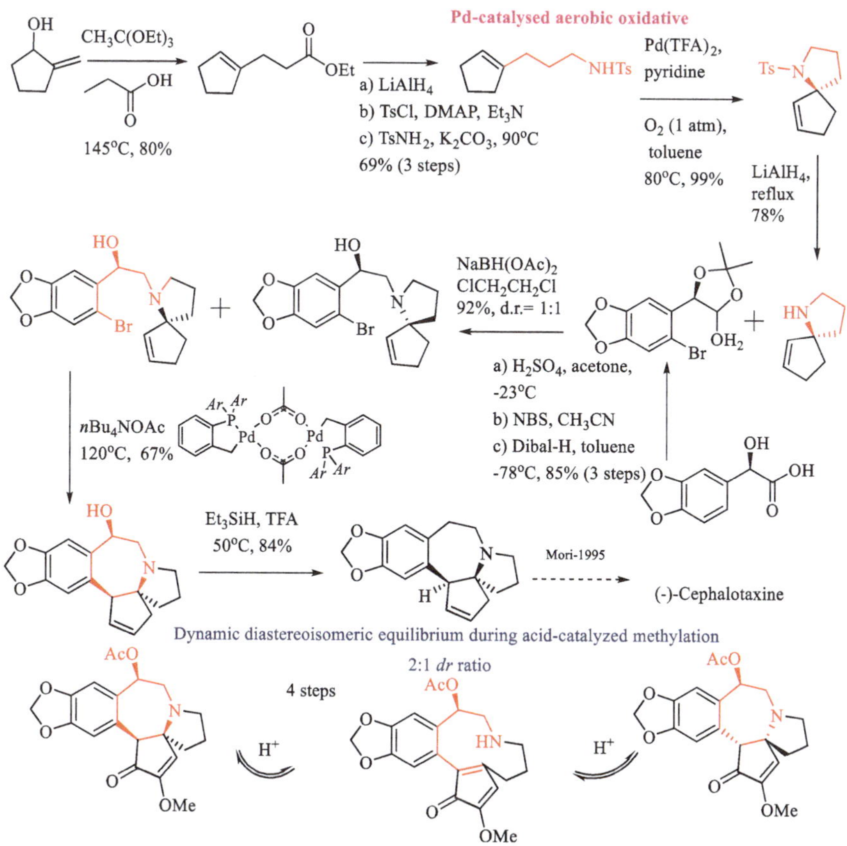


SCHEME 38 Synthesis via Pd-catalyzed aerobic oxidative heterocyclization chemistry.[30]

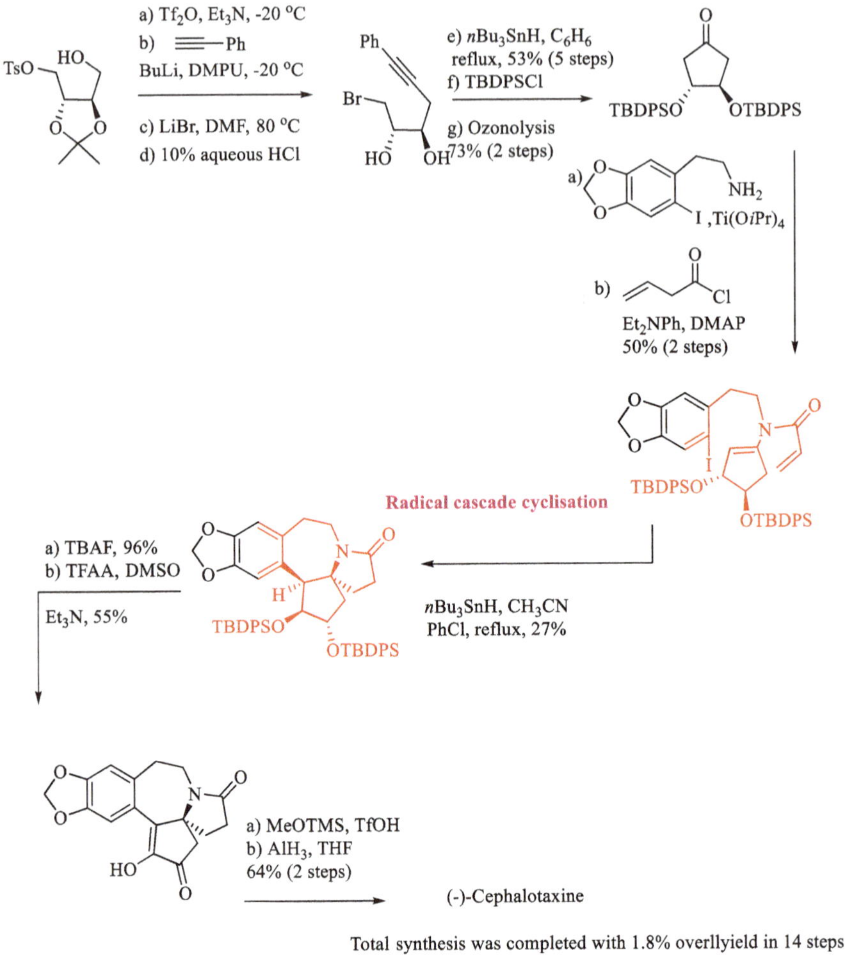


SCHEME 39 Total synthesis via radical cascade cyclization reactions.[19]

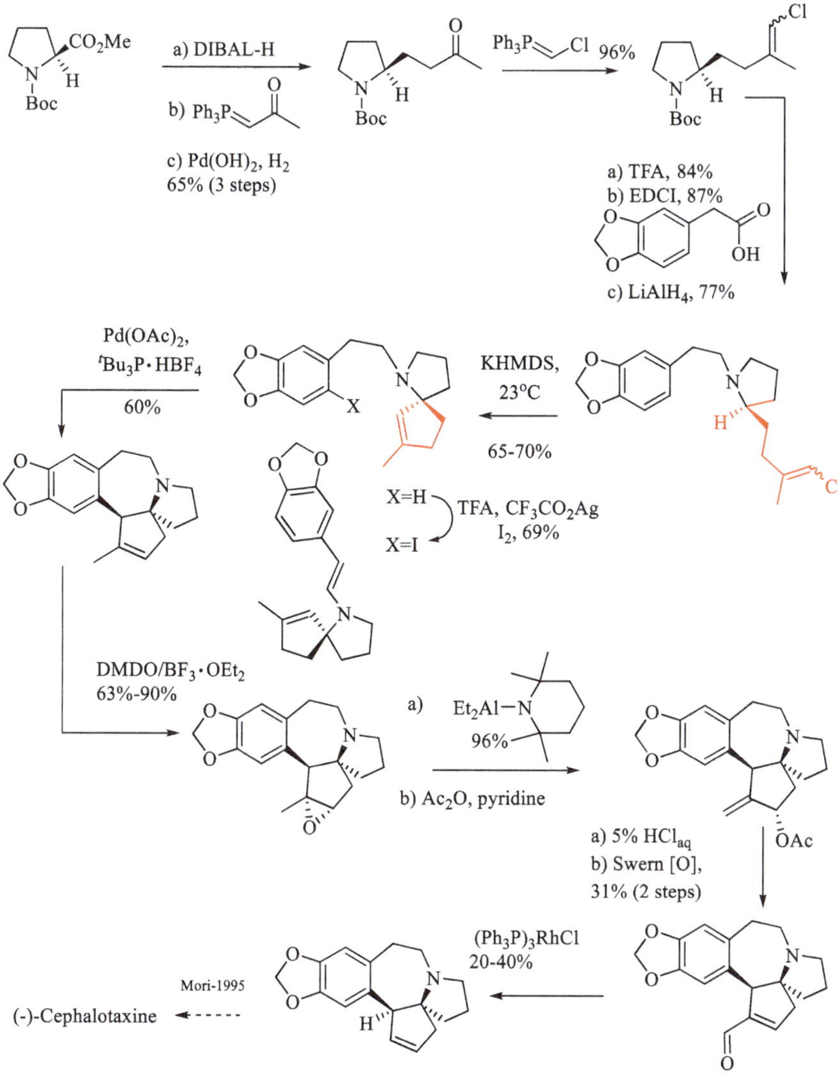


**SCHEME 40** Formal synthesis via 1,5-CH insertion reaction.[31]

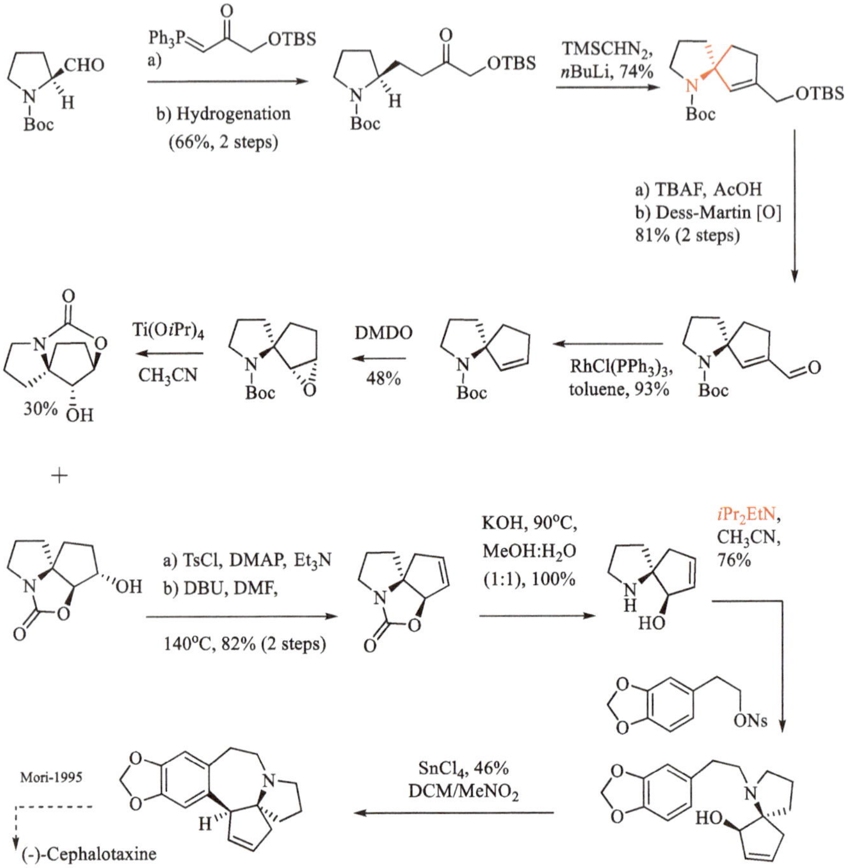


SCHEME 41 Formal synthesis via 1,5-CH insertion reaction.[32]

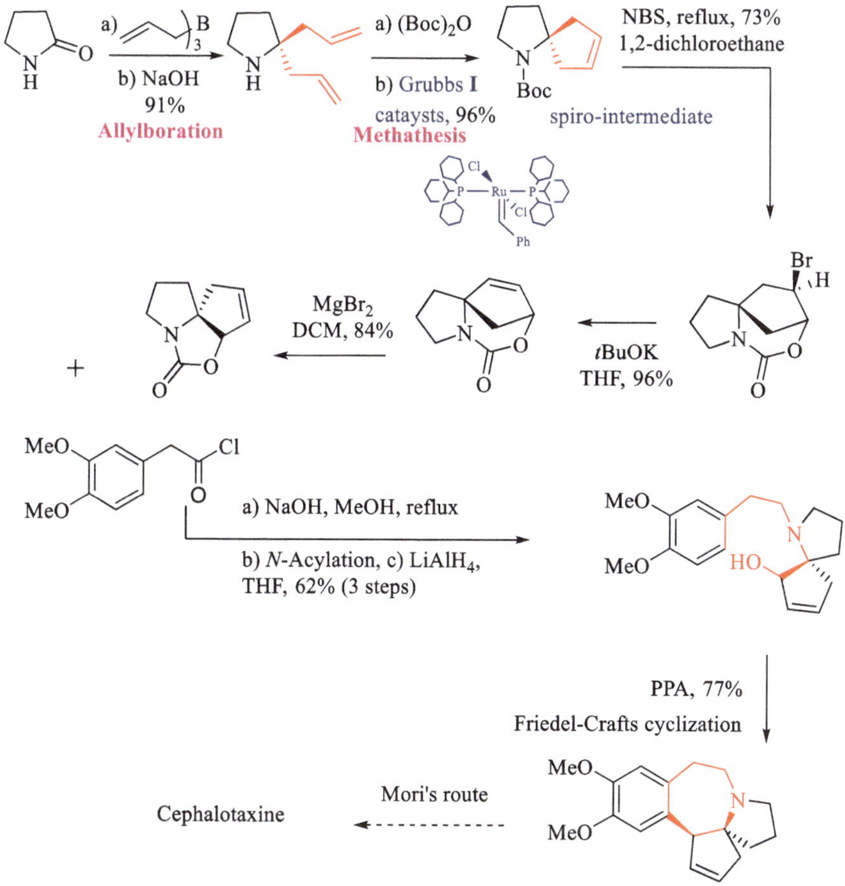


**SCHEME 42** Approach toward cephalotaxine via Friedel–Crafts cyclization.

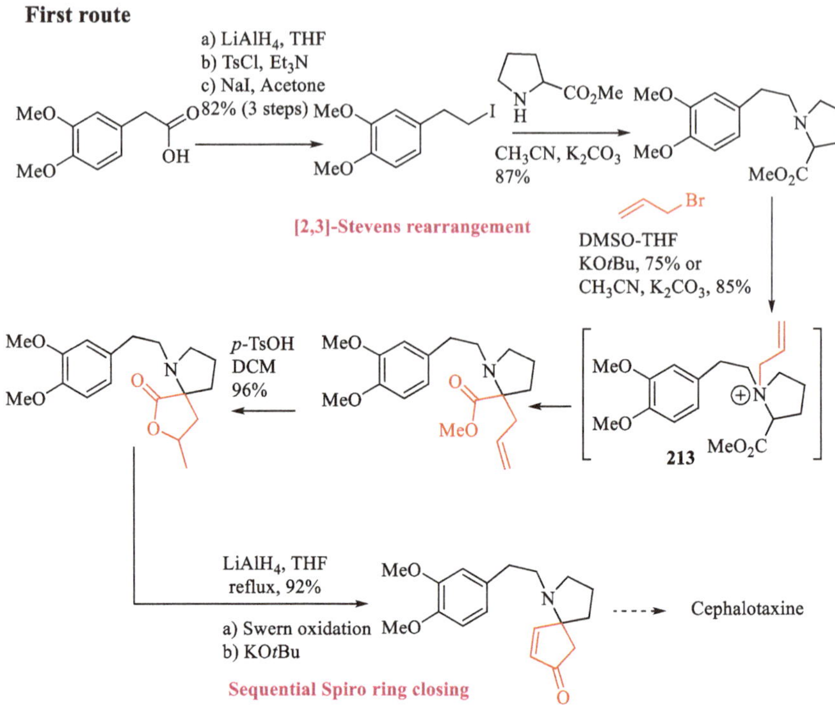


SCHEME 43 Formal synthesis via [2,3]-Stevens rearrangement—acid lactonization.[34]

## Second route

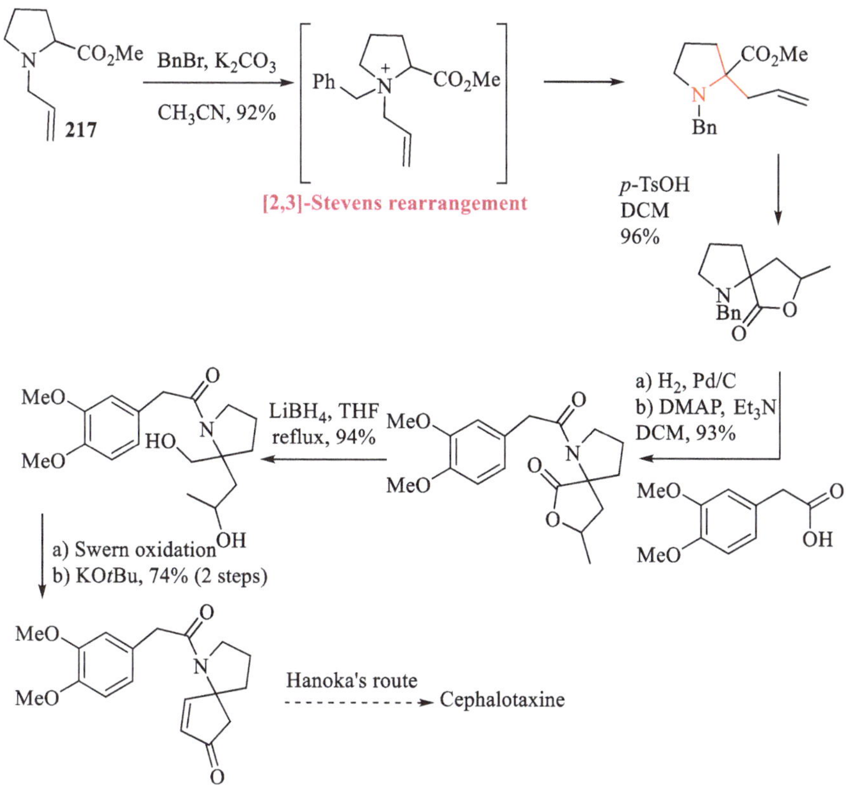


SCHEME 44 Second synthetic route.

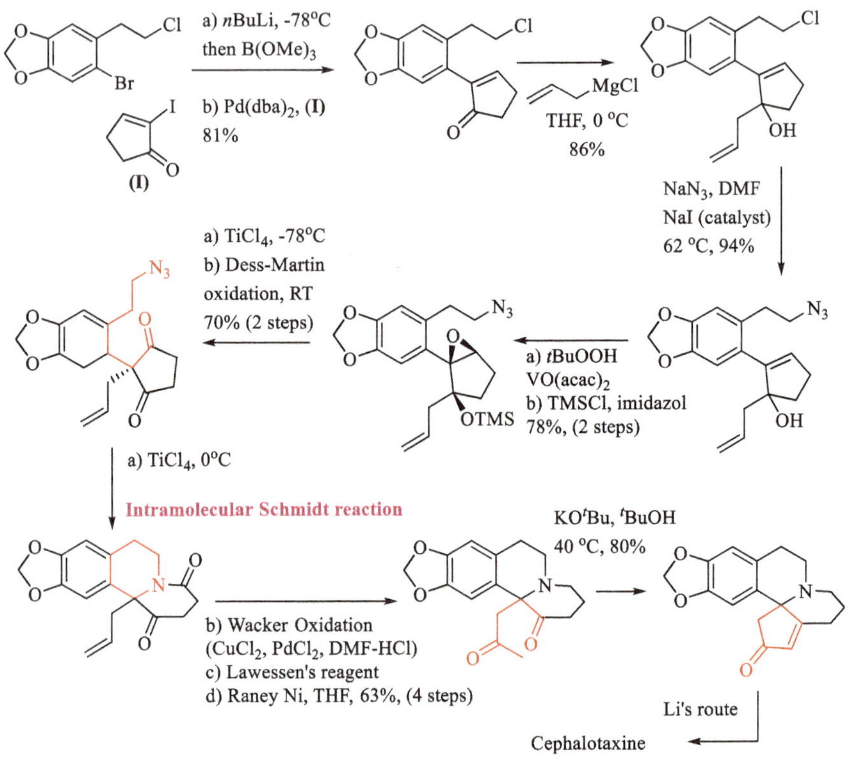


**SCHEME 45** Synthesis via intermolecular Schmidt reaction.[35]

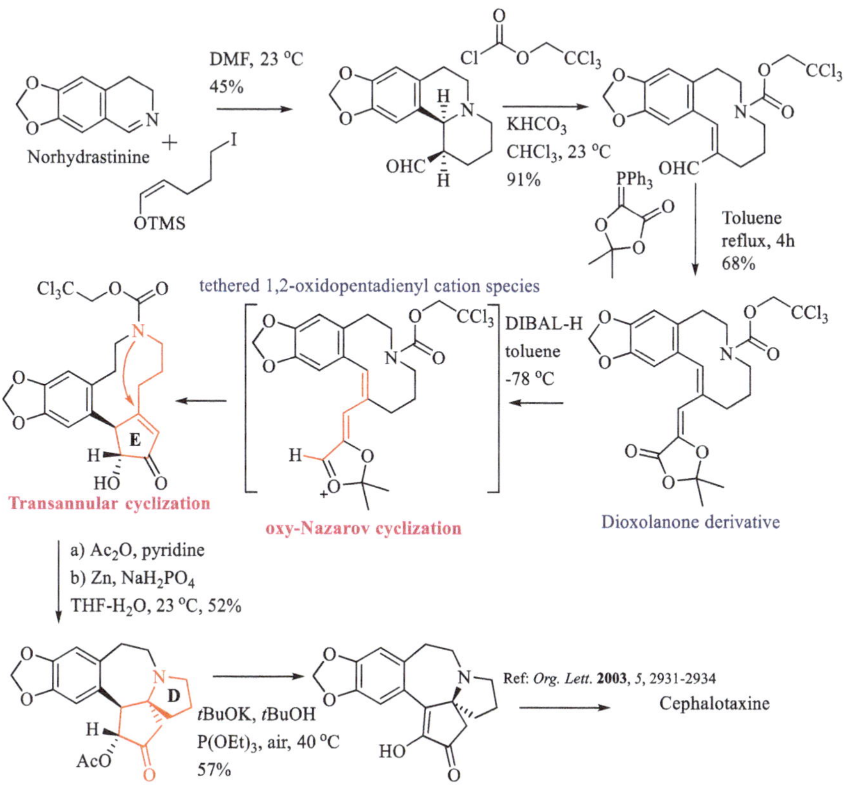


**SCHEME 46** Total synthesis via oxy-Nazarov cyclization and transannulation strategies.[36]

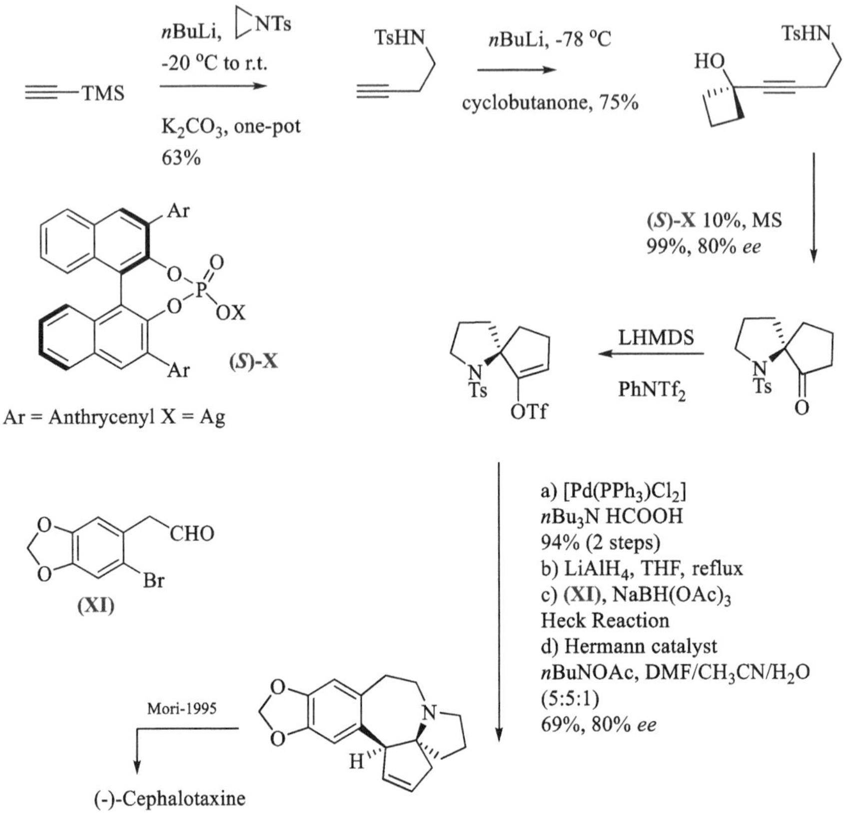


**SCHEME 47** (−)-Formal synthesis via tandem hydroamination/semipinacol rearrangement reaction.[37]

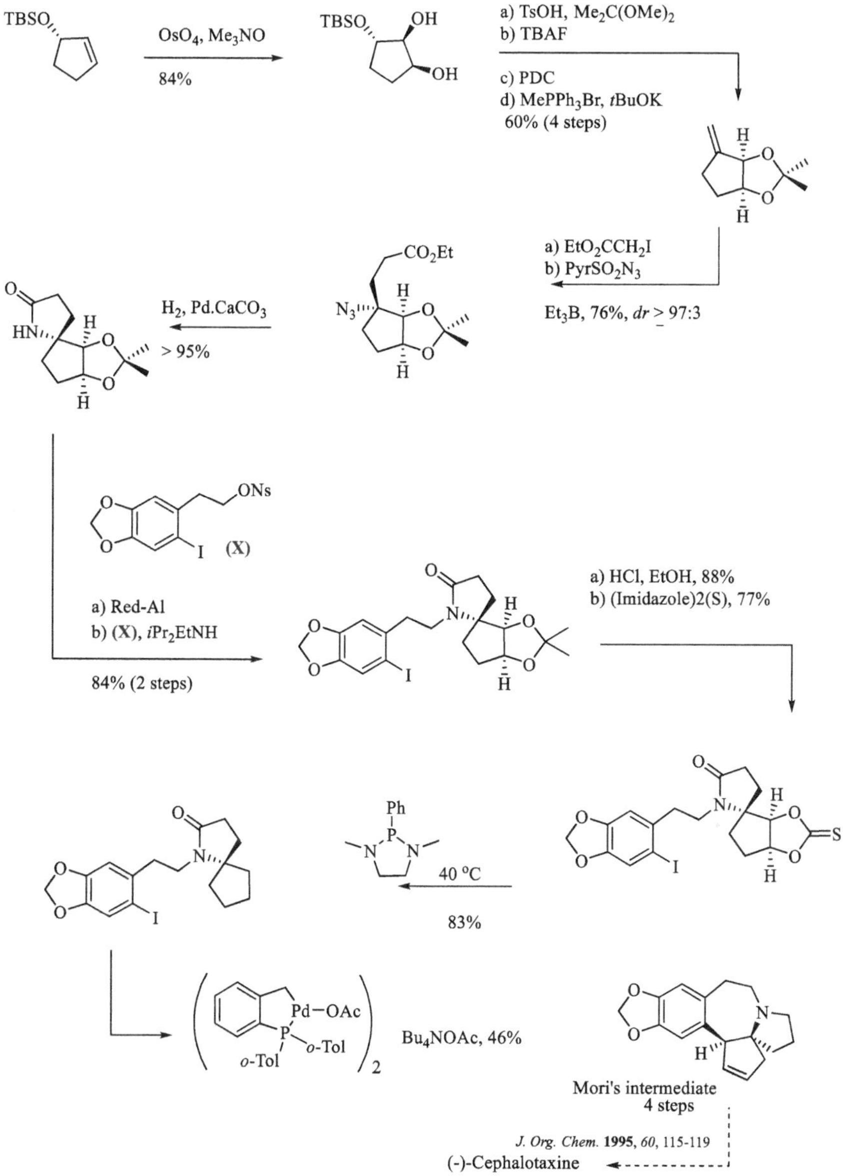


**SCHEME 48** (−)-Formal synthesis via stereoselective radical carboazidation.[38]

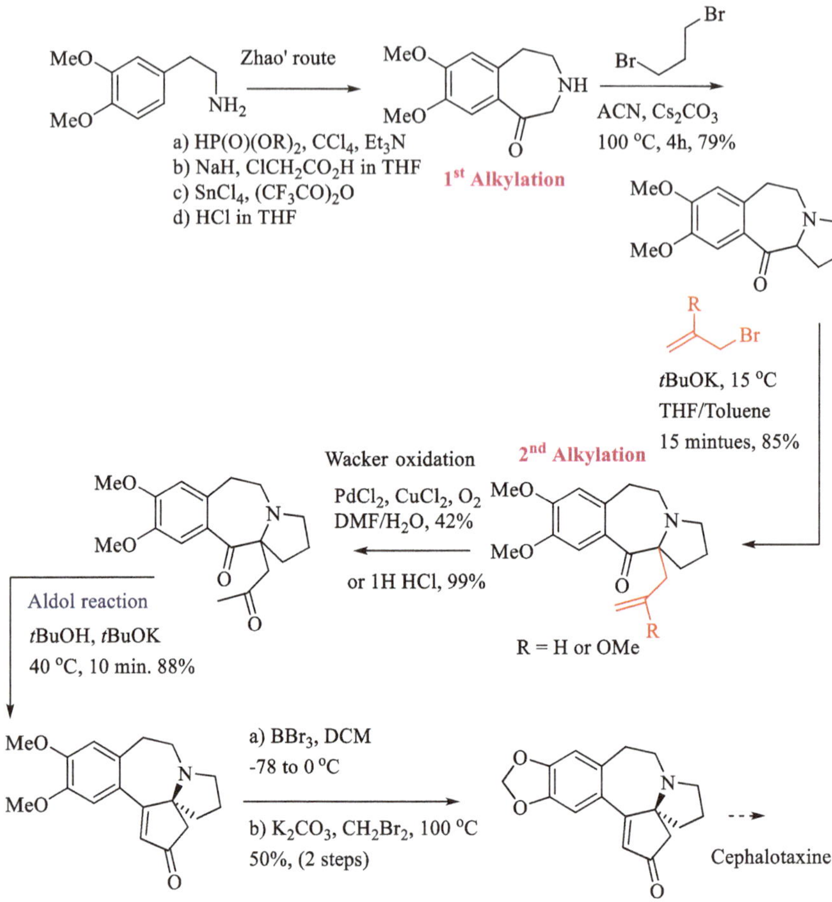


SCHEME 49 Formal synthesis of cephalotaxine via double alkylation.[39]

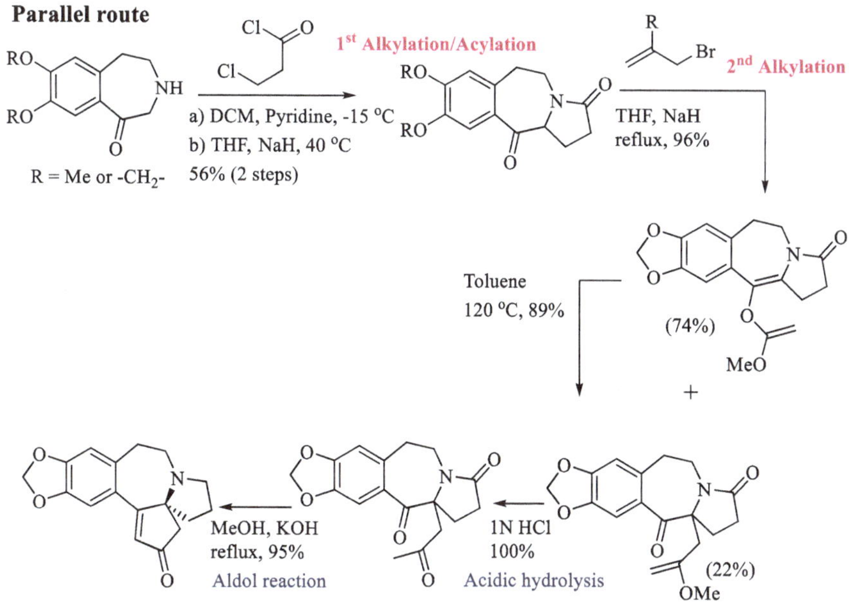


SCHEME 50 Parallel synthetic route.

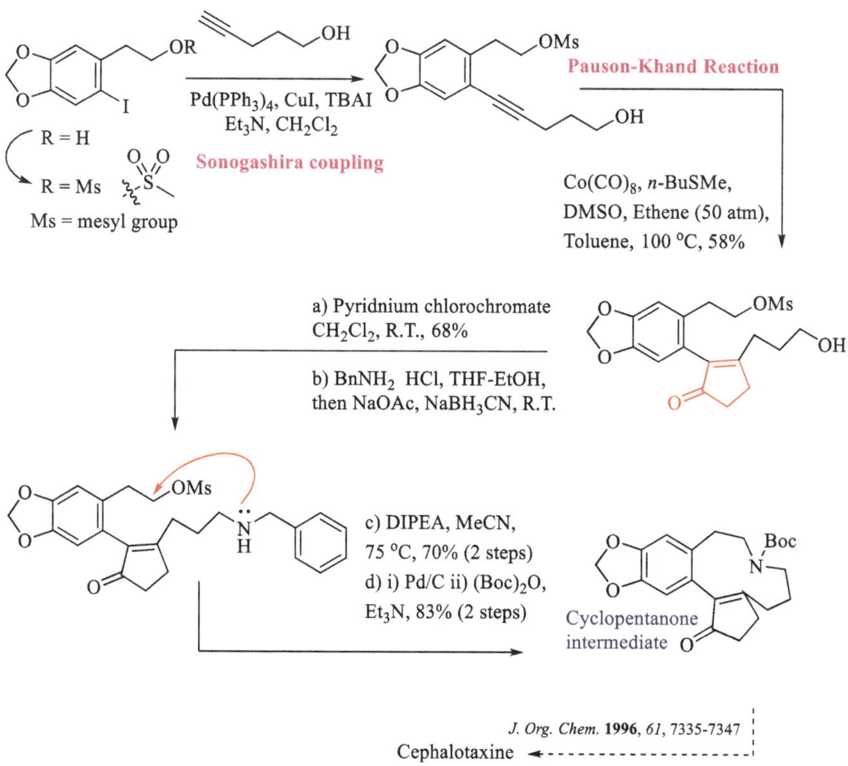


SCHEME 51 Formal synthesis via Pauson-Khand reaction.[40]

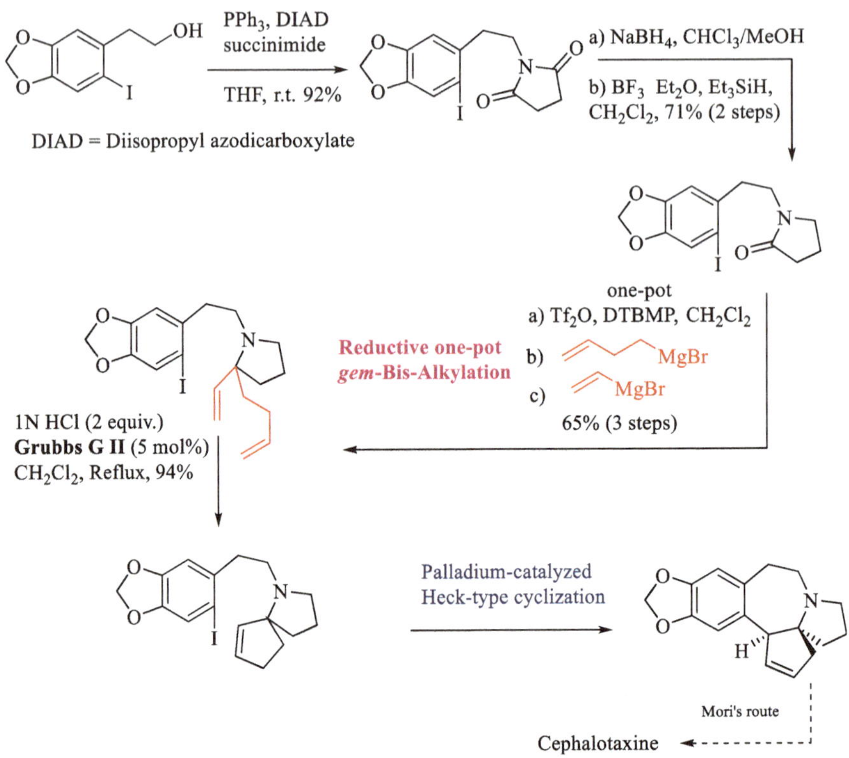


SCHEME 52 Formal synthesis via reductive gem-bis-alkylation.[41]

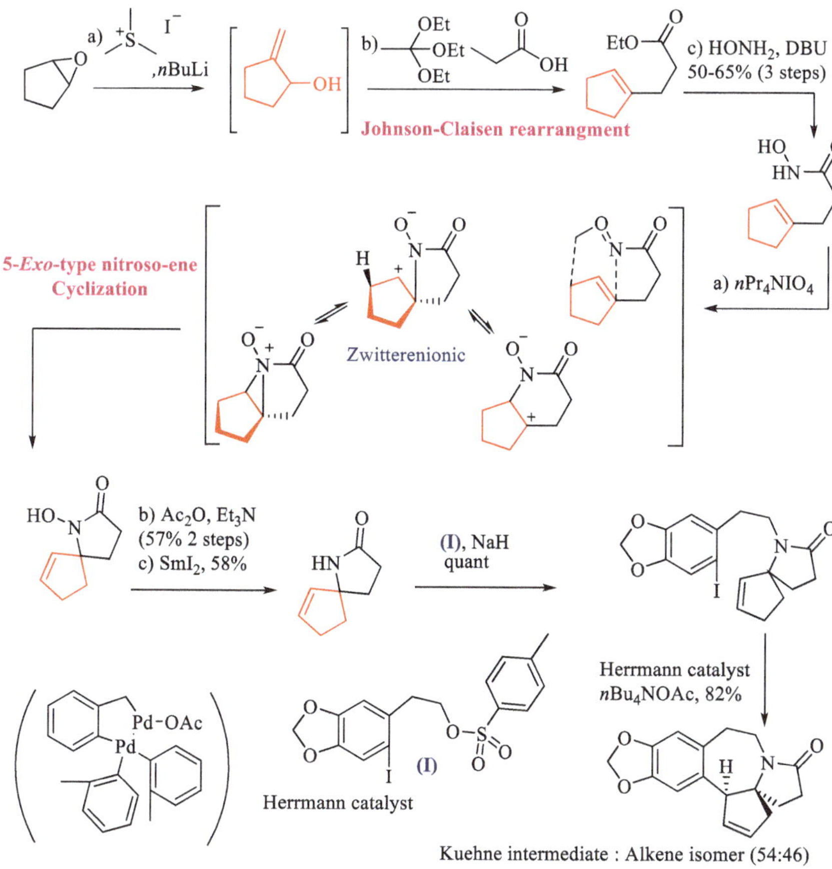


**SCHEME 53** Formal synthesis via nitroso-ene cyclization.[42]

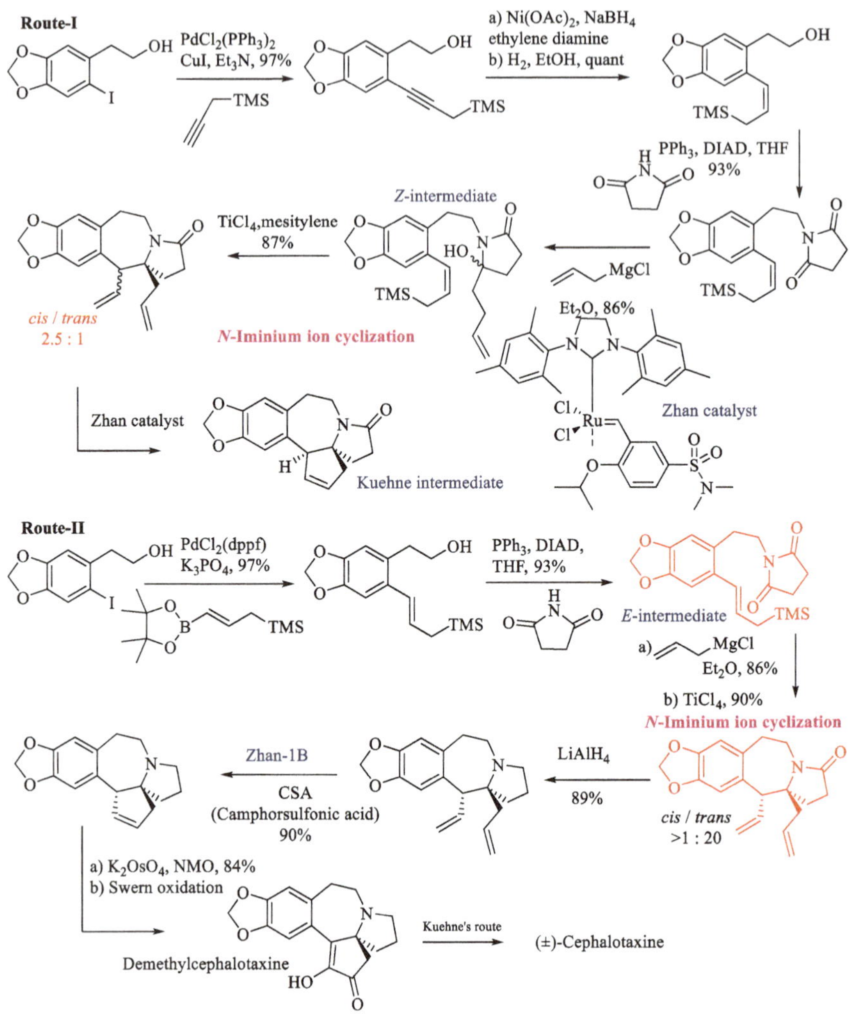


SCHEME 54 Formal synthesis via N-iminium ion cyclization.[43]

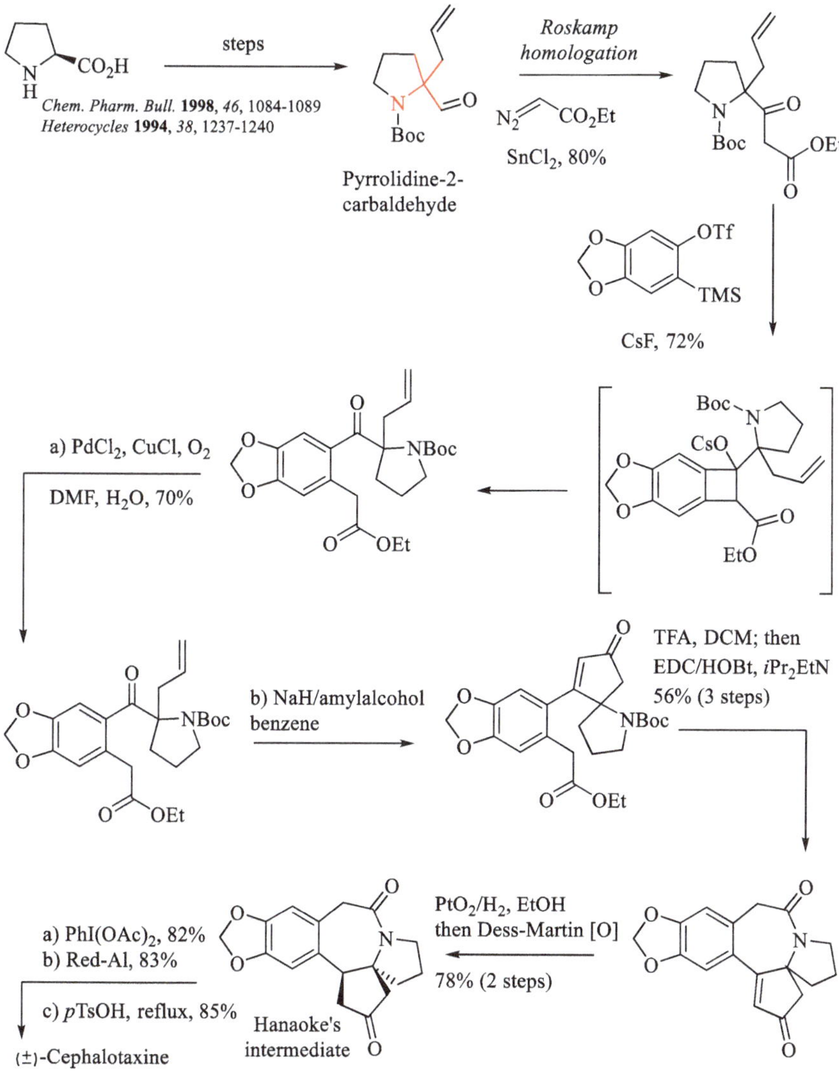


Total synthesis was completed with 10% overall yield in 9 steps from pyrrolidine-2-carbaldehyde

**SCHEME 55** Formal total synthesis via Aryne insertion reaction.[44]

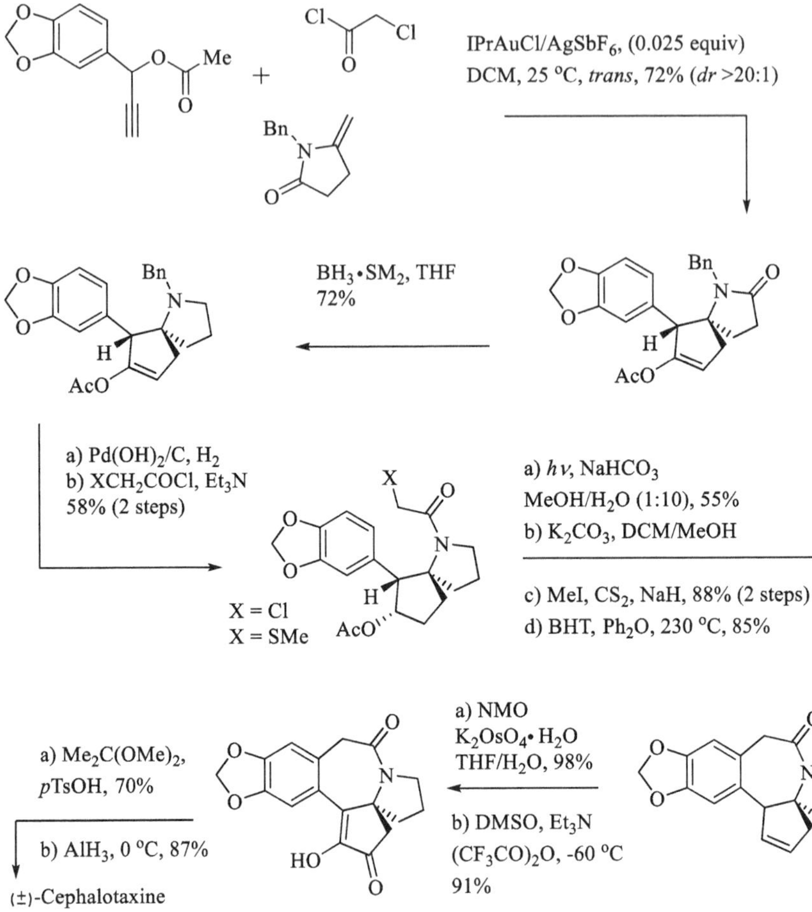


**SCHEME 56** Total synthesis via Au-catalyzed [2 + 3] annulation reaction.[45]

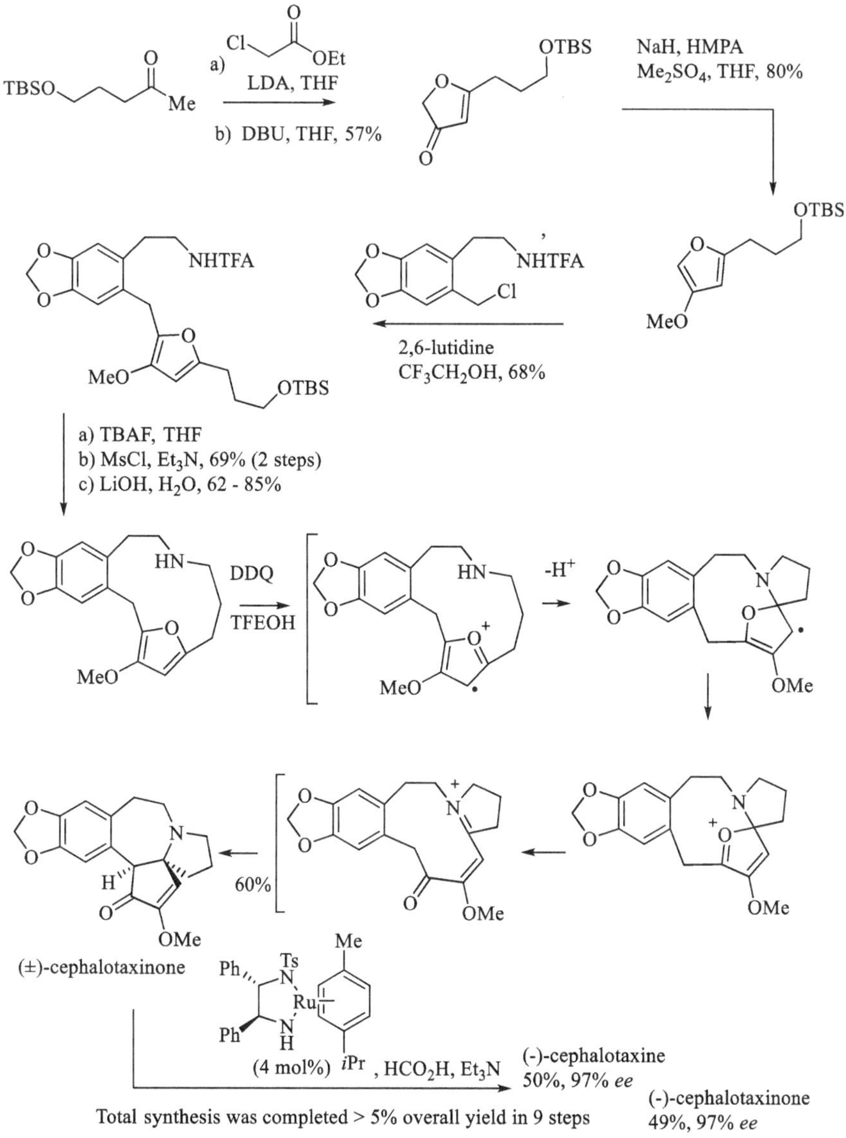


**SCHEME 57** Synthesis via furan oxidation–transannular Mannich cyclization.[46]

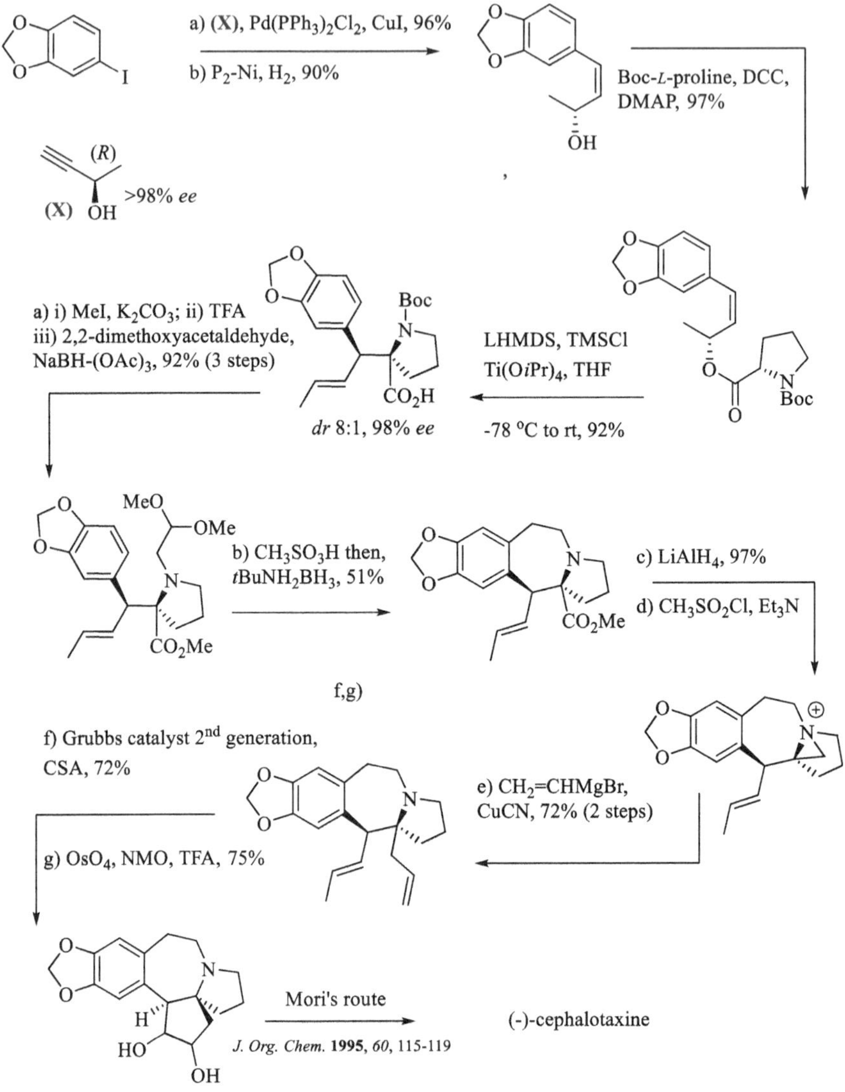


**SCHEME 58** Formal (−)-total synthesis via proline ester enolate Claisen rearrangement (EECR).[47]

# References


1. (a) Jalil Miah, M. A.; Hudlicky, T.; Reed, J. W. In *The Alkaloids,* Vol., Cordell, G. A., Ed.; Acadmic Press: San Diego, 1998; pp. 199–269.
   (b) Huang, L.; Xue, X. In *The Alkaloids,* Vol. 23, Brossi, A., Ed.; Academic Press: New York, 1984.
2. Powell, R.; Weisleder, D.; Smith, C., Jr. Antitumor Alkaloids from Cephalotaxus Harringtonia: Structure and Activity. *J. Pharmacol. Sci.* **1972,** *61,* 1227–1230.
3. Paudler, W. W.; Kerley, G. I.; McKay, J. The Alkaloids of Cephalotaxus Drupacea and Cephalotaxus Fortunei. *J. Org. Chem.* **1963,** *28,* 2194–2197.
4. Powell, R.; Weisleder, D.; Smith, C.; Wolff, I.; Haylett, T.; Swart, L. Structure of Cephalotaxine and Related Alkaloids. *Tetrahedron Lett.* **1969,** *46,* 4081–4084, 2679101.
5. Arora, S.; Bates, R.; Grady, R.; Powell, R. Crystal and Molecular Structure of Cephalotaxine P-Bromobenzoate. *J. Org. Chem.* **1974,** *39,* 1269–1271.
6. Parry, R. J.; Chang, M. N. T.; Schwab, J. M.; Foxman, B. M. Biosynthesis of the *Cephalotaxus* Alkaloids. Investigations of the Early and Late Stages of Cephalotaxine Biosynthesis. *J. Am. Chem. Soc.* **1980,** *102,* 1099–1111.
7. Weinreb, S. M.; Auerbach, J. Total Synthesis of the *Cephalotaxus* Alkaloids. Cephalotaxine, Cephalotaxinone, and Demethylcephalotaxinone. *J. Am. Chem. Soc.* **1975,** *97,* 2503.
8. Semmelhack, M. F.; Chong, B. P.; Stauffer, R. D.; Rogerson, T. D.; Chong, A.; Jones, L. D. Total Synthesis of the *Cephalotaxus* Alkaloids. Problem in Nucleophilic Aromatic Substitution. *J. Am. Chem. Soc.* **1975,** *97,* 2507–2516.
9. Yasuda, S.; Yamada, T.; Hanaoka, M. A Novel and Stereoselective Synthesis of (±)-cephalotaxine and its Analogue. *Tetrahedron Lett.* **1986,** *27,* 2023–2026.
10. Yasuda, S.; Yamamoto, Y.; Yoshida, S.; Hanaoka, M. A Total Synthesis of (±)-cephalotaxinamide. *Chem. Pharm. Bull.* **1988,** *36,* 4229–4231.
11. Kuehne, M. E.; Bornmann, W. G.; Parsons, W. H.; Spitzer, T. D.; Blount, J. F.; Zubieta, J. J. Total Syntheses of (±)-Cephalotaxine and (±)-8-Oxocephalotaxin. *J. Am. Chem. Soc.* **1988,** *53,* 3439–3450.
12. Burkholder, T.; Fuchs, P. Total Synthesis of *Dl*-Cephalotaxine: the First Example of an Intramolecular 4+2 Cycloaddition where the Dienophile Has Been Delivered from the Face Opposite to the Tethering Moiety. *J. Am. Chem. Soc.* **1988,** *110,* 2341–2342.
13. (a) Ishibashi, H.; Okano, M.; Tamaki, H.; Maruyama, K.; Yakura, T.; Ikeda, M. Total Synthesis of (±)-cephalotaxine. *J. Chem. Soc., Chem. Commun.* **1990,** *20,* 1436–1437.
    (b) Ikeda, M.; Okano, M.; Kosaka, K.; Kido, M.; Ishibashi, H. Synthetic Studies on Cephalotaxus Alkaloids. A Synthesis of (±)-cephalotaxine. *Chem. Pharm. Bull.* **1993,** *41,* 276–281.
14. Seebach, D.; Boes, M.; Naef, R.; Schweizer, W. B. Alkylation of Amino Acids without Loss of the Optical Activity: Preparation of. alpha.-substituted Proline Derivatives. A Case of Self-Reproduction of Chirality. *J. Am. Chem. Soc.* **1983,** *105,* 5390–5398.
15. Isono, N.; Mori, M. Total Synthesis of (-)-cephalotaxine. *J. Org. Chem.* **1995,** *60,* 115–119.
16. a) Lin, X.; Kavash, R. W.; Mariano, P. S. Two Interrelated Strategies for Cephalotaxine Synthesis. *J. Org. Chem.* **1996,** *61,* 7335–7347.
    b) Lin, X.; Kavash, R. W.; Mariano, P. S. A Cephalotaxine Synthesis Founded on a Mechanistically Interesting, Quasi-Biomimetic Strategy. *J. Am. Chem. Soc.* **1994,** *116,* 9791–9792.
17. a) Lin, X.; Kavash, R. W.; Mariano, P. S. Two Interrelated Strategies for Cephalotaxine Synthesis. *J. Org. Chem.* **1996,** *61,* 7335–7347.
    b) Lin, X.; Kavash, R. W.; Mariano, P. S. A Cephalotaxine Synthesis Founded on a Mechanistically Interesting, Quasi-Biomimetic Strategy. *J. Am. Chem. Soc.* **1994,** *116,* 9791–9792.

18. Nagasaka, T.; Sato, H.; Saeki, S.-i. 1-Azaspiro [4.4] Nonane-2, 6-dione and the Separation and Absolute Configurations of its Enantiomers. *Tetrahedron: Asymmetry* **1997,** *8,* 191–194.
19. Ikeda, M.; El Bialy, S. A.; Hirose, K.-i.; Kotake, M.; Bayomi, S. M.; Shehata, I. A.; Abdelal, A. M.; Yakura, T. A Formal Total Synthesis of (-)-cephalotaxine. *Chem. Pharm. Bull.* **1999,** *47,* 983–987.
20. Tietze, L. F.; Schirok, H. Enantioselective Highly Efficient Synthesis of (-)-cephalotaxine Using Two Palladium-Catalyzed Transformations. *J. Am. Chem. Soc.* **1999,** *121,* 10264–10269.
21. Herrmann, W. A.; Brossmer, C.; Reisinger, C. P.; Riermeier, T. H.; Öfele, K.; Beller, M. Palladacycles: Efficient New Catalysts for the Heck Vinylation of Aryl Halides. *Chem. Eur J.* **1997,** *3,* 1357–1364.
22. Suga, S.; Watanabe, M.; Yoshida, J. Electroauxiliary-assisted Sequential Introduction of Two Carbon Nucleophiles on the Same Alpha-Carbon of Nitrogen: Application to the Synthesis of Spiro Compounds. *J. Am. Chem. Soc.* **2002,** *124,* 14824–14825.
23. Li, W. D.; Wang, Y. Q. A Novel and Efficient Total Synthesis of Cephalotaxine. *Org. Lett.* **2003,** *5,* 2931–2934.
24. Planas, L.; Pérard-Viret, J.; Royer, J. Stereoselective Synthesis of (−)-Cephalotaxine and C-7 Alkylated Analogues. *J. Org. Chem.* **2004,** *69,* 3087–3092.
25. Li, W. D.; Ma, B. C. A Novel Formal Total Synthesis of Cephalotaxine. *J. Org. Chem.* **2005,** *70,* 3277–3280.
26. Ma, B. C.; Wang, Y. Q.; Li, W. D. An alternative synthesis of Dolby-Weinreb enamine en route to cephalotaxine. *J. Org. Chem.* **2005,** *70,* 4528–4530.
27. Zhao, Z.; Mariano, P. S. Application of the Photocyclization Reaction of 1, 2-Cyclopenta-Fused Pyridinium Perchlorate to Formal Total Syntheses of (−)-cephalotaxine. *Tetrahedron* **2006,** *62,* 7266–7273.
28. Eckelbarger, J. D.; Wilmot, J. T.; Gin, D. Y. Strain-release Rearrangement of *N*-Vinyl-2-Arylaziridines. Total Synthesis of the Anti-leukemia Alkaloid (−)-deoxyharringtonine. *J. Am. Chem. Soc.* **2006,** *128,* 10370–10371.
29. Li, W. D.; Wang, X. W. Novel Formal Synthesis of Cephalotaxine *via* a Facile Friedel-Crafts Cyclization. *Org. Lett.* **2007,** *9,* 1211–1214.
30. Liu, Q.; Ferreira, E. M.; Stoltz, B. M. Convergency and Divergency as Strategic Elements in Total Synthesis: The Total Synthesis of (−)-Drupacine and the Formal Total Synthesis of (±)-Cephalotaxine, (−)-Cephalotaxine, and (+)-Cephalotaxine. *J. Org. Chem.* **2007,** *72* (19), 7352–7358.
31. Esmieu, W. R.; Worden, S. M.; Catterick, D.; Wilson, C.; Hayes, C. J. A Formal Synthesis of (-)-cephalotaxine. *Org. Lett.* **2008,** *10,* 3045–3048.
32. Hameed, A.; Blake, A. J.; Hayes, C. J. A Second Generation Formal Synthesis of (−)-cephalotaxine. *J. Org. Chem.* **2008,** *73,* 8045–8048.
33. Kuznetsov, N. Y.; Kolomnikova, G. D.; Khrustalev, V. N.; Golovanov, D. G.; Bubnov, Y. N. The Combination of Diallylboration and Ring-Closing Metathesis in the Synthesis of Spiro-β-Amino Alcohols and (±)-Cephalotaxine. *Eur. J. Org Chem.* **2008,** *2008,* 5647–5655.
34. Sun, M.-r.; Lu, H.-t.; Wang, Y.-z.; Yang, H.; Liu, H.-m. Highly Efficient Formal Synthesis of Cephalotaxine, Using the Stevens Rearrangement–Acid Lactonization Sequence as A Key Transformation. *J. Org. Chem.* **2009,** *74,* 2213–2216.
35. Zhao, Y. M.; Gu, P.; Zhang, H. J.; Zhang, Q. W.; Fan, C. A.; Tu, Y. Q.; Zhang, F. M. Formal Syntheses of (+/-)-stemonamine and (+/-)-cephalotaxine. *J. Org. Chem.* **2009,** *74,* 3211–3213.
36. Li, W.-D. Z.; Duo, W.-G.; Zhuang, C.-H. Concise Total Synthesis of (±)-Cephalotaxine *via* a Transannulation Strategy: Development of a Facile Reductive Oxy-Nazarov Cyclization. *Org. Lett.* **2011,** *13,* 3538–3541.

37. Zhang, Q. W.; Xiang, K.; Tu, Y. Q.; Zhang, S. Y.; Zhang, X. M.; Zhao, Y. M.; Zhang, T. C. Formal Synthesis of (−)-Cephalotaxine Based on a Tandem Hydroamination/Semipinacol Rearrangement Reaction. *Chem. Asian J.* **2012,** *7,* 894–898.
38. Gonçalves-Martin, M. G.; Zigmantas, S.; Renaud, P. Formal Synthesis of (−)-Cephalotaxine. *Helv. Chim. Acta* **2012,** *95,* 2502–2514.
39. Zhang, Z.-W.; Zhang, X.-F.; Feng, J.; Yang, Y.-H.; Wang, C.-C.; Feng, J.-C.; Liu, S. Formal Synthesis of Cephalotaxine. *J. Org. Chem.* **2012,** *78,* 786–790.
40. Xing, P.; Huang, Z.-g.; Jin, Y.; Jiang, B. A Formal Synthesis of (±)-Cephalotaxine via Pauson–Khand Reaction. *Synthesis* **2013,** *45,* 596–600.
41. Xiao, K. J.; Luo, J. M.; Xia, X. E.; Wang, Y.; Huang, P. Q. General One-Pot Reductive gem-Bis-Alkylation of Tertiary Lactams/Amides: Rapid Construction of 1-Azaspirocycles and Formal Total Synthesis of (±)-Cephalotaxine. *Chem. Eur J.* **2013,** *19,* 13075–13086.
42. Huang, S.-H.; Tian, X.; Mi, X.; Wang, Y.; Hong, R. Nitroso-*ene* Cyclization Enabled Access to 1-azaspiro [4.4] Nonane and its Application in a Modular Synthesis toward (±)-cephalotaxine. *Tetrahedron Lett.* **2015,** *56,* 6656–6658.
43. Liu, H.; Yu, J.; Li, X.; Yan, R.; Xiao, J.-C.; Hong, R. Stereoselectivity in N-Iminium Ion Cyclization: Development of an Efficient Synthesis of (±)-Cephalotaxine. *Org. Lett.* **2015,** *17,* 4444–4447.
44. Gouthami, P.; Chegondi, R.; Chandrasekhar, S. Formal Total Synthesis of (±)-Cephalotaxine and Congeners *via* Aryne Insertion Reaction. *Org. Lett.* **2016,** *18,* 2044–2046.
45. Ma, X.-Y.; An, X.-T.; Zhao, X.-H.; Du, J.-Y.; Deng, Y.-H.; Zhang, X.-Z.; Fan, C.-A. Au-Catalyzed [2+3] Annulation of Enamides with Propargyl Esters: Total Synthesis of Cephalotaxine and Cephalezomine H. *Org. Lett.* **2017,** *19,* 2965–2968.
46. Ju, X.; Beaudry, C. M. Total Synthesis of (−)-Cephalotaxine and (−)-Homoharringtonine *via* Furan Oxidation–Transannular Mannich Cyclization. *Angew. Chem. Int. Ed.* **2019,** *58,* 6752–6755.
47. Jeon, H.; Chung, Y.; Kim, S. Proline Ester Enolate Claisen Rearrangement and Formal Total Synthesis of (−)-Cephalotaxine. *J. Org. Chem.* **2019,** *84,* 8080–8089.

CHAPTER

# 5

# Kaitocephalin

## 5.1 Abstract

Kaitocephalin, a novel pyrrolidine-based amino acid, was isolated by Shin-ya et al. in 1997 from the filamentous fungus *Eupenicillium shearii*.[1] The structural framework of the molecules showed the conjugation of three amino acids include alanine, proline, and serine via carbon–carbon linkages. The natural product has challenging synthetic features that include the construction of central trisubstituted proline core possessing *N*-bearing quaternary center at C4 position. Initially, the stereochemistry of natural product was proposed (2*S*,3*S*,4*R*,7*R*,9*S*) on the basis of NMR data [2];[2] however, the structure was later revised to (2*R*,3*S*,4*R*,7*R*,9*S*) based on synthetic studies [3].[3] The biological evaluation of kaitocephalin showed strong antagonist activity against 2-amino-3-(3-hydroxy-5-methylisoxazole-4-ylpropionate (AMPA), kainic acid (KA)[4] [4], and *N*-methyl-D-aspartate (NMDA) glutamate receptors, which have key role in various physiological processes such as learning, memory, and neural plasticity [5].[5] Thus, the pharmaceutically important antagonist can serve as lead for the development of new therapeutic agent to treat neurological diseases. The fascinating structural framework and significantly biological potential encourage and drag the attention of chemists in the relevant area to synthesize molecule for further processes followed to develop new drug molecules (Fig. 5).

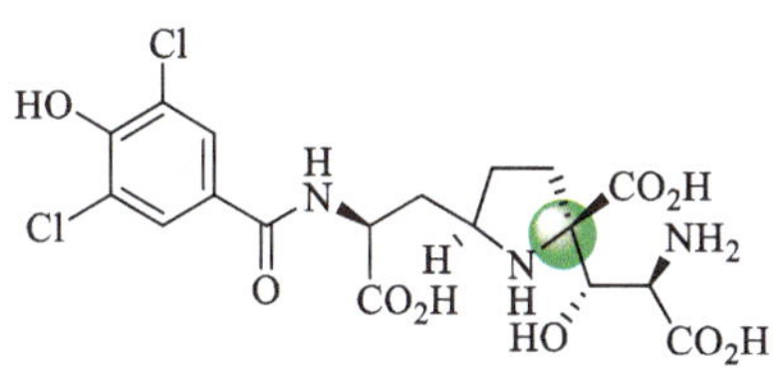


(-) Kaitocephalin

**FIGURE 5** Structures of kaitocephalin.


https://doi.org/10.1016/B978-0-12-822262-1.00005-6

## 5.2 Kitahara's total synthesis (2002)

See (Scheme 59).

## 5.3 Kitahara's total synthesis (2002)

See (Scheme 60).

## 5.4 Ohfune's total enantioselective synthesis (2005)

See (Scheme 61).

## 5.5 Chamberlin's total synthesis (2008)

See (Scheme 62).

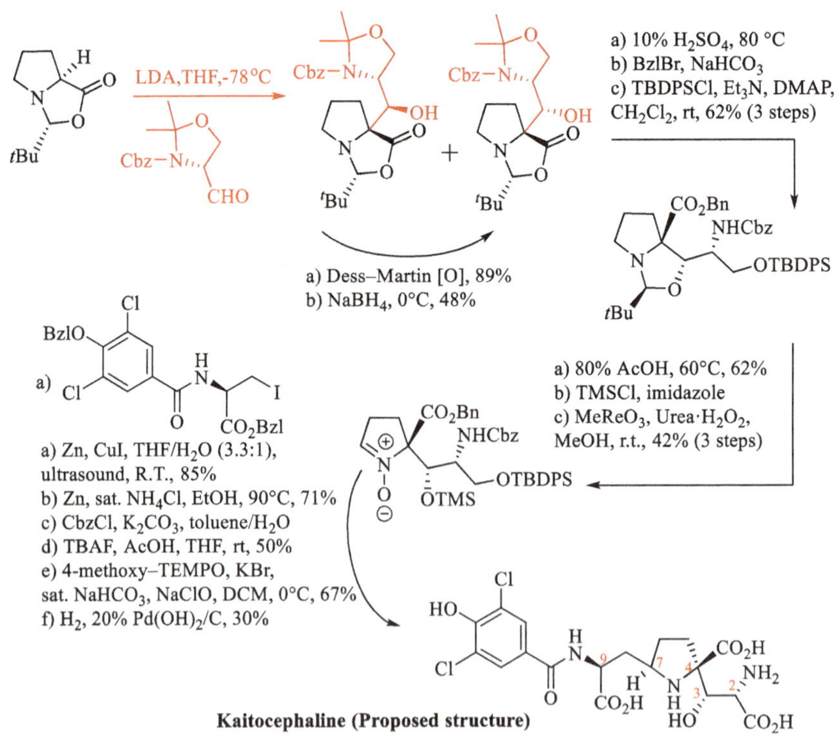


SCHEME 59 Total synthesis of proposed structure.[3]

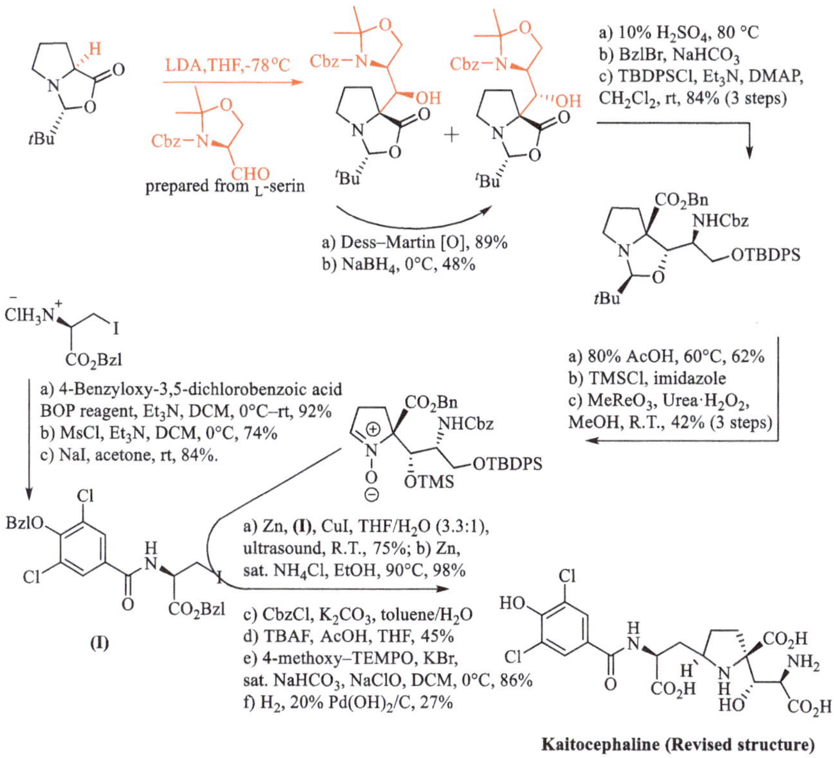


SCHEME 60 Total synthesis of revised structure.[6]

## 5.6 Ohfune's total enantioselective synthesis (2009)

See (Scheme 63).

## 5.7 Ma's reinvestigation of kaitocephalin (2011)

See (Scheme 64).

## 5.8 Hatakeyama's total synthesis (2012)

See (Scheme 65).

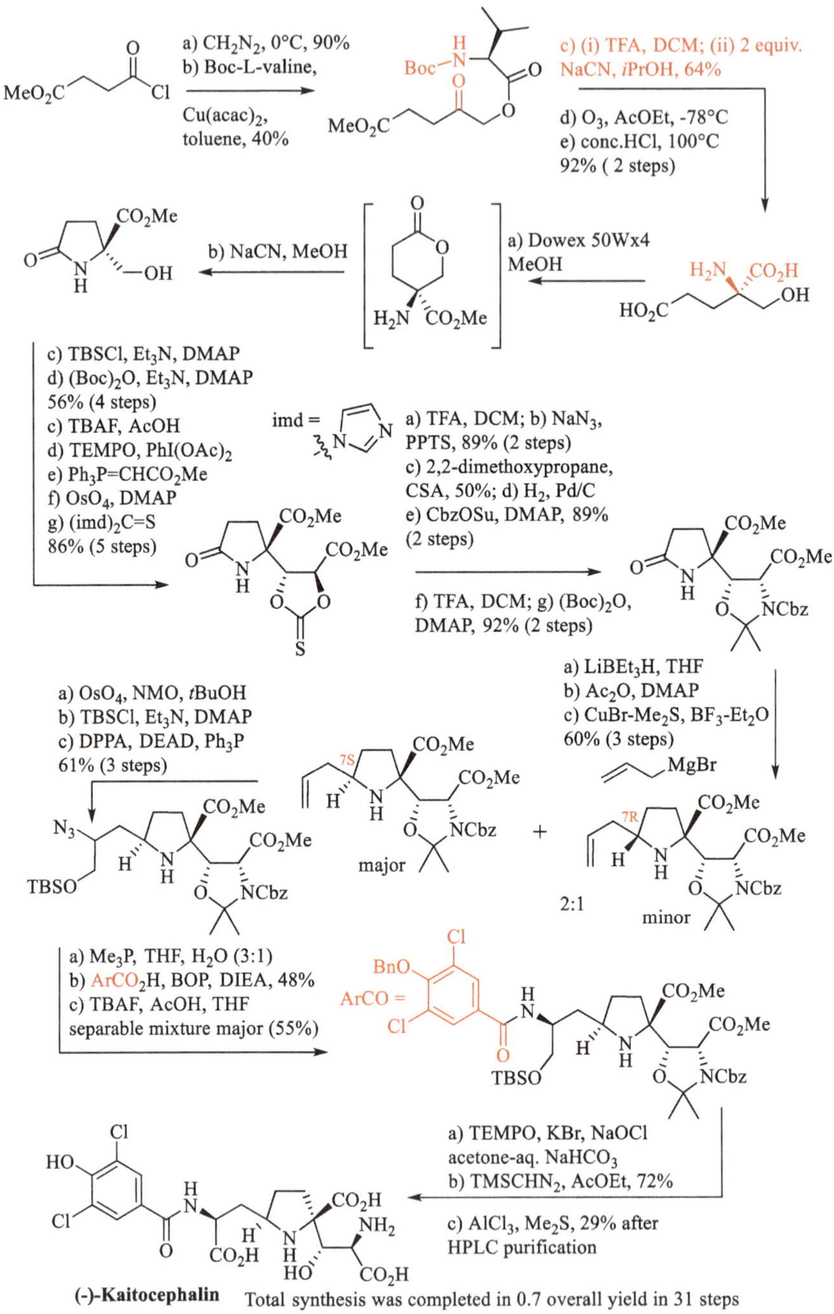


**SCHEME 61** Total enantioselective synthesis.[7]

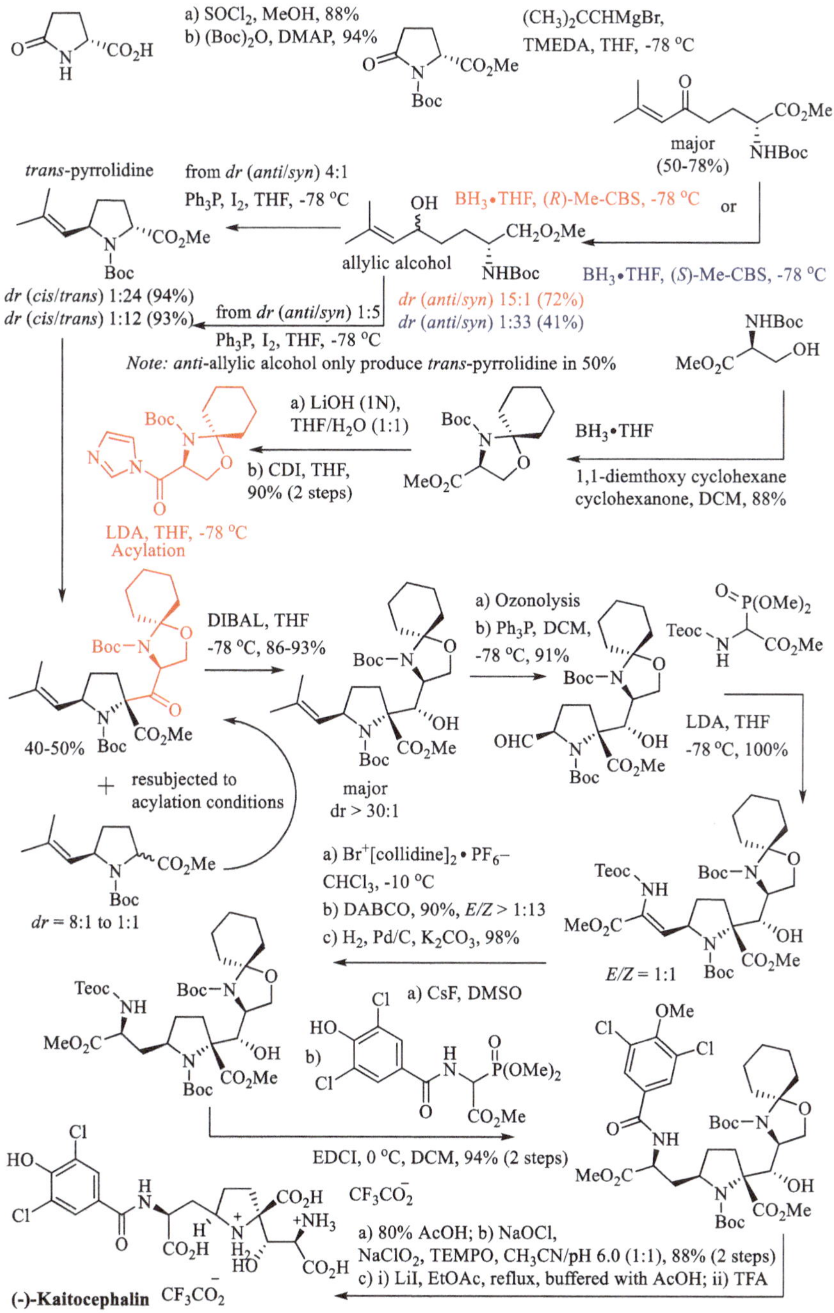


SCHEME 62 Stereocontrolled total synthesis.[8]

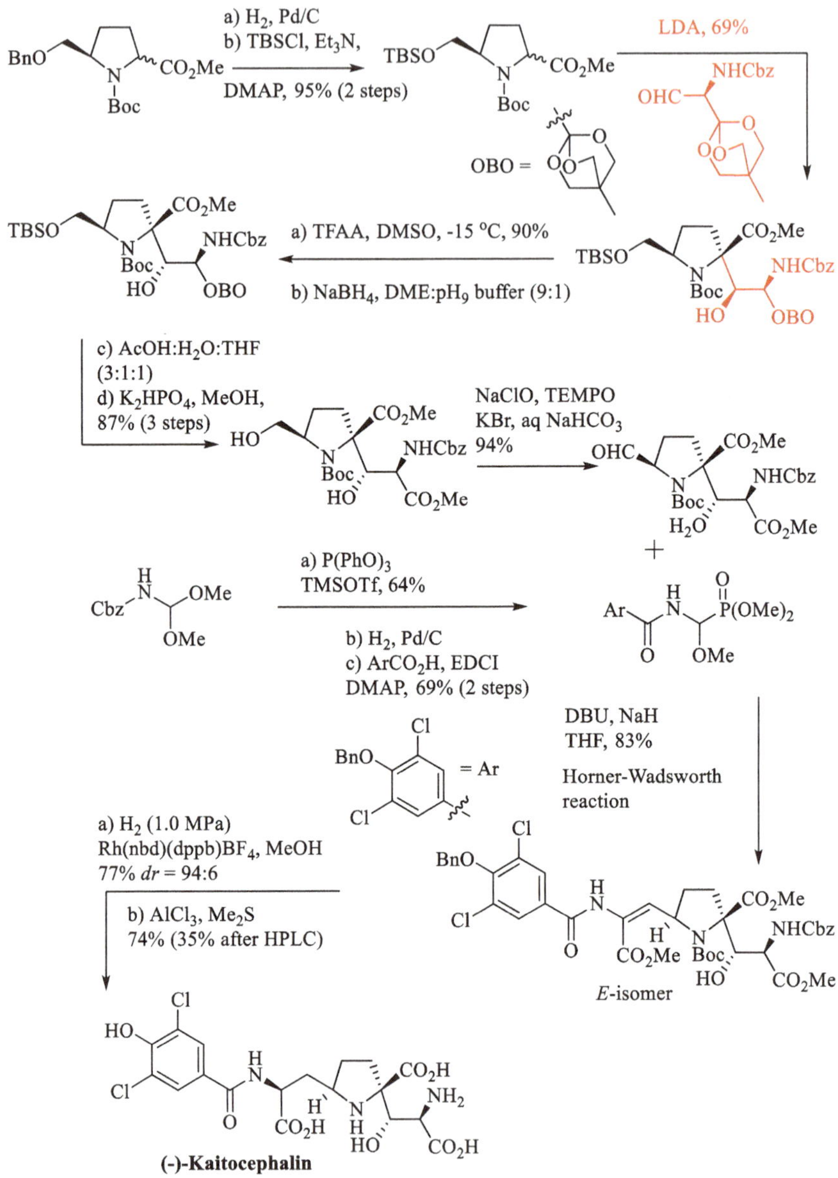


SCHEME 63 Total enantioselective synthesis.[9]

## 5.9 Kang's kaitocephalin total synthesis (2013)

See (Scheme 66).

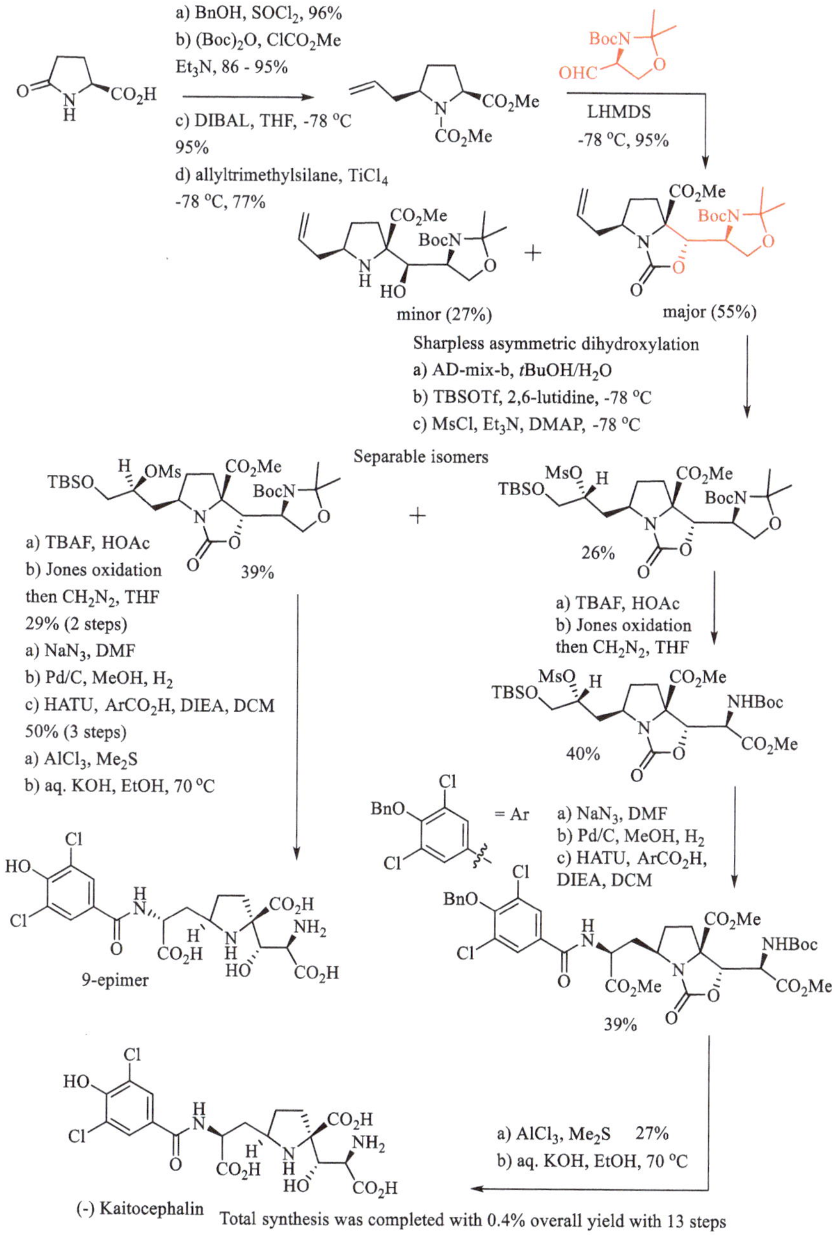


**SCHEME 64** Reinvestigation of kaitocephalin.[10]

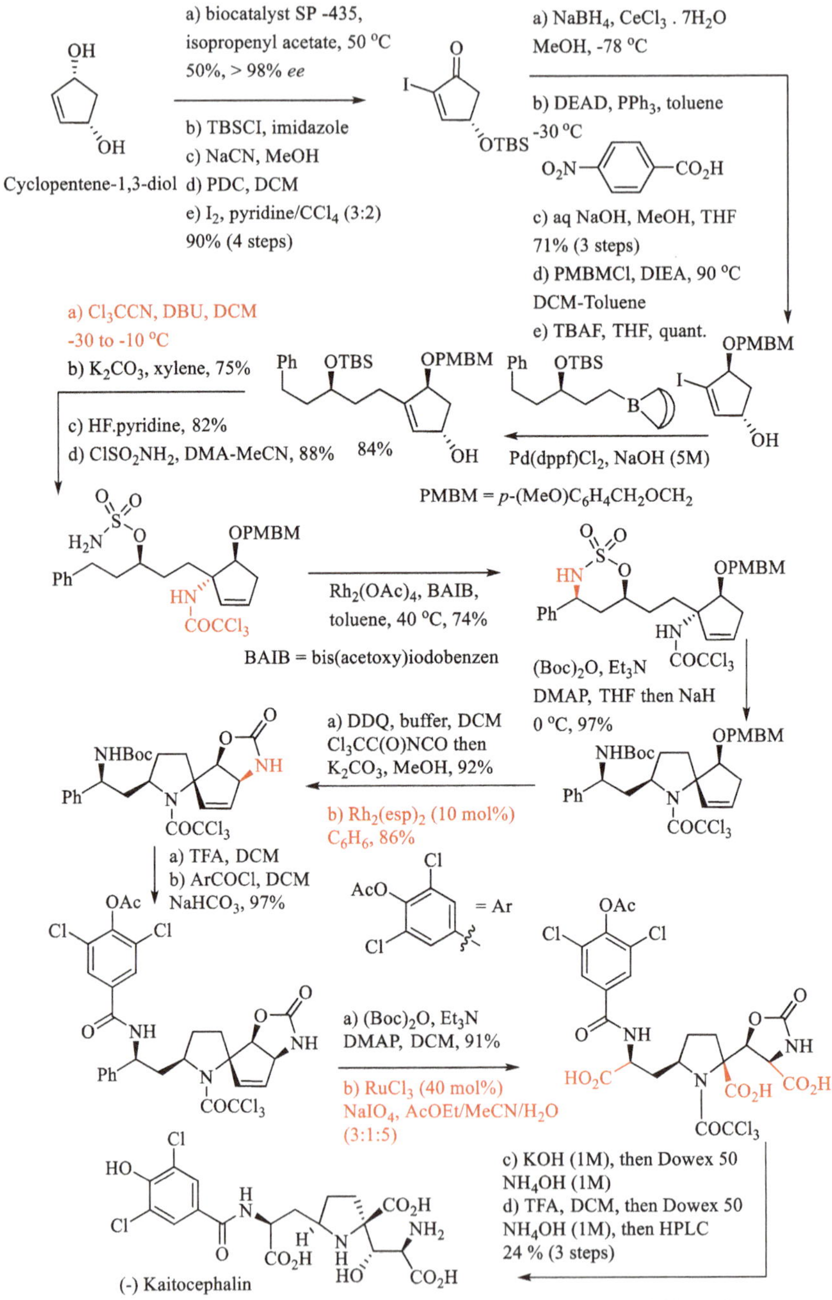


Total synthesis was completed with 1.8 % overall yield with 25 steps .

**SCHEME 65** Synthesis *via* Rh-catalyzed CH amination.[11]

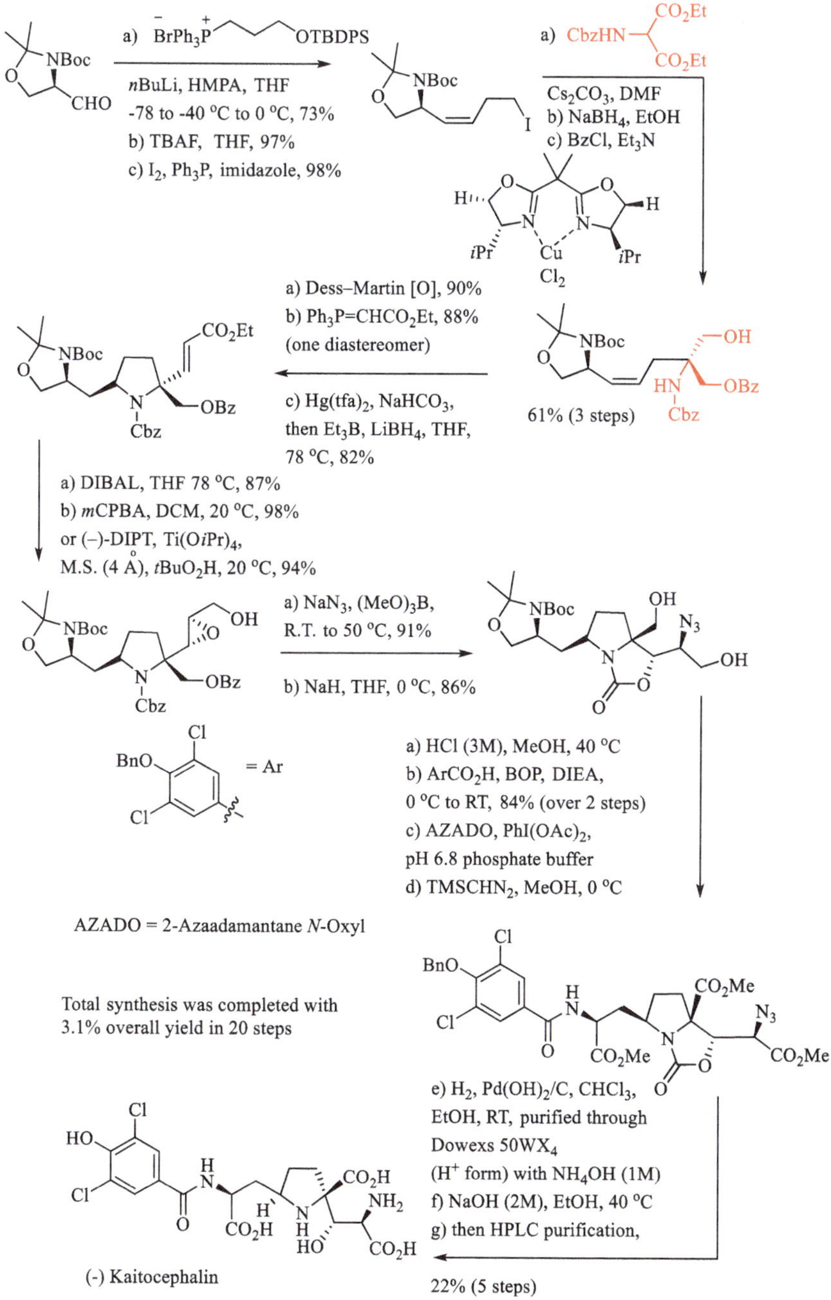


**SCHEME 66** Kaitocephalin total synthesis.[12]

## 5.10 Garner's synthesis *via* [C + NC + C] coupling (2014)

See (Scheme 67).

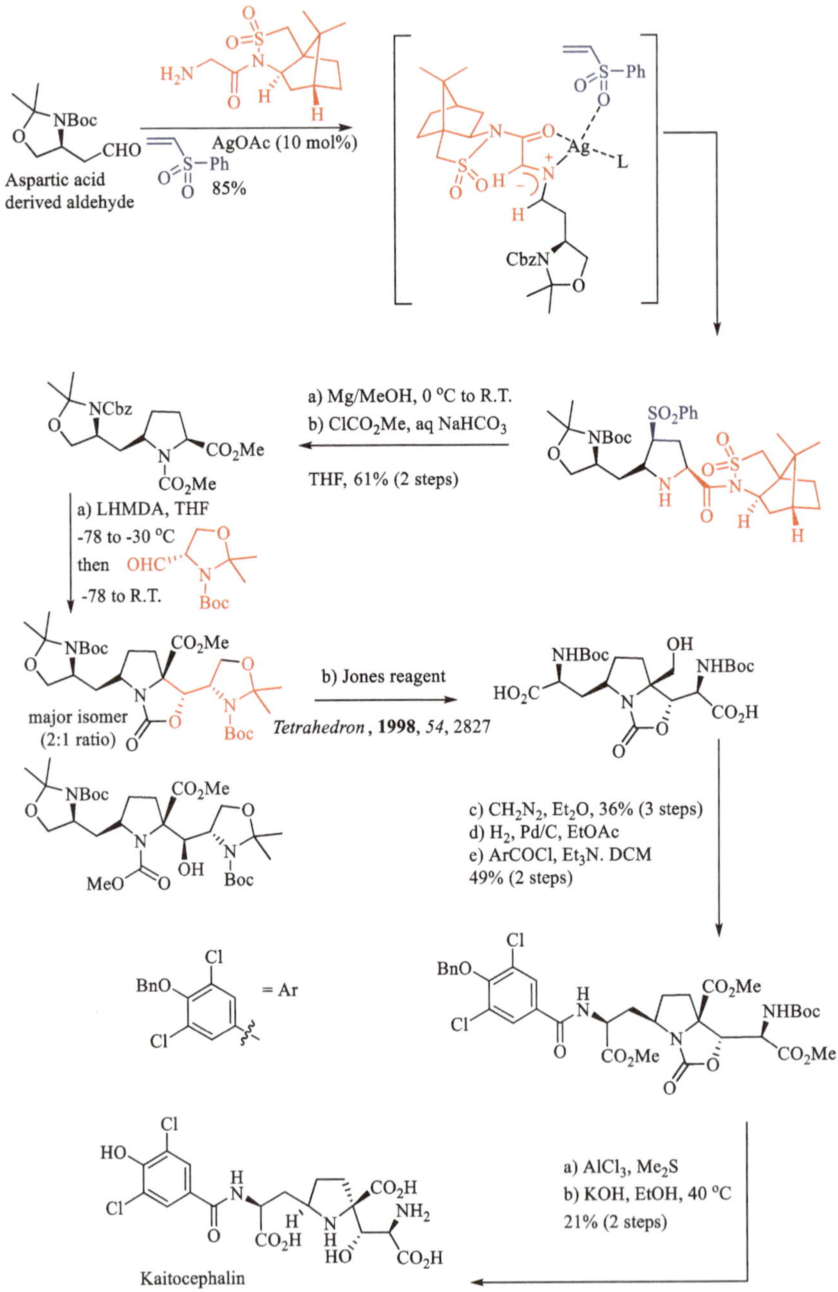


**SCHEME 67** Synthesis *via* [C + NC + C] coupling.[13]

## 5.11 Dhavale's formal synthesis (2014)

See (Scheme 68).

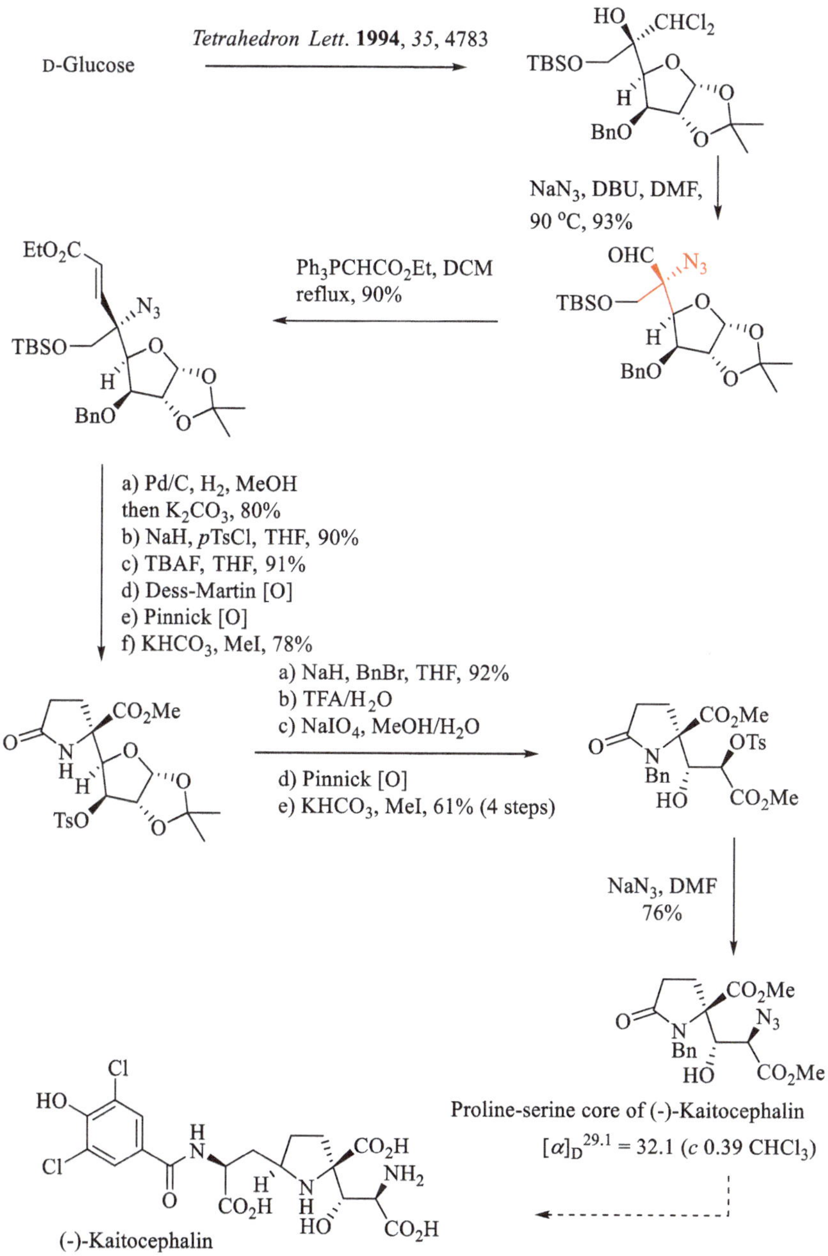


SCHEME 68 Formal synthesis.[14]

## 5.12 Lee's total synthesis (2019)

See (Scheme 69).

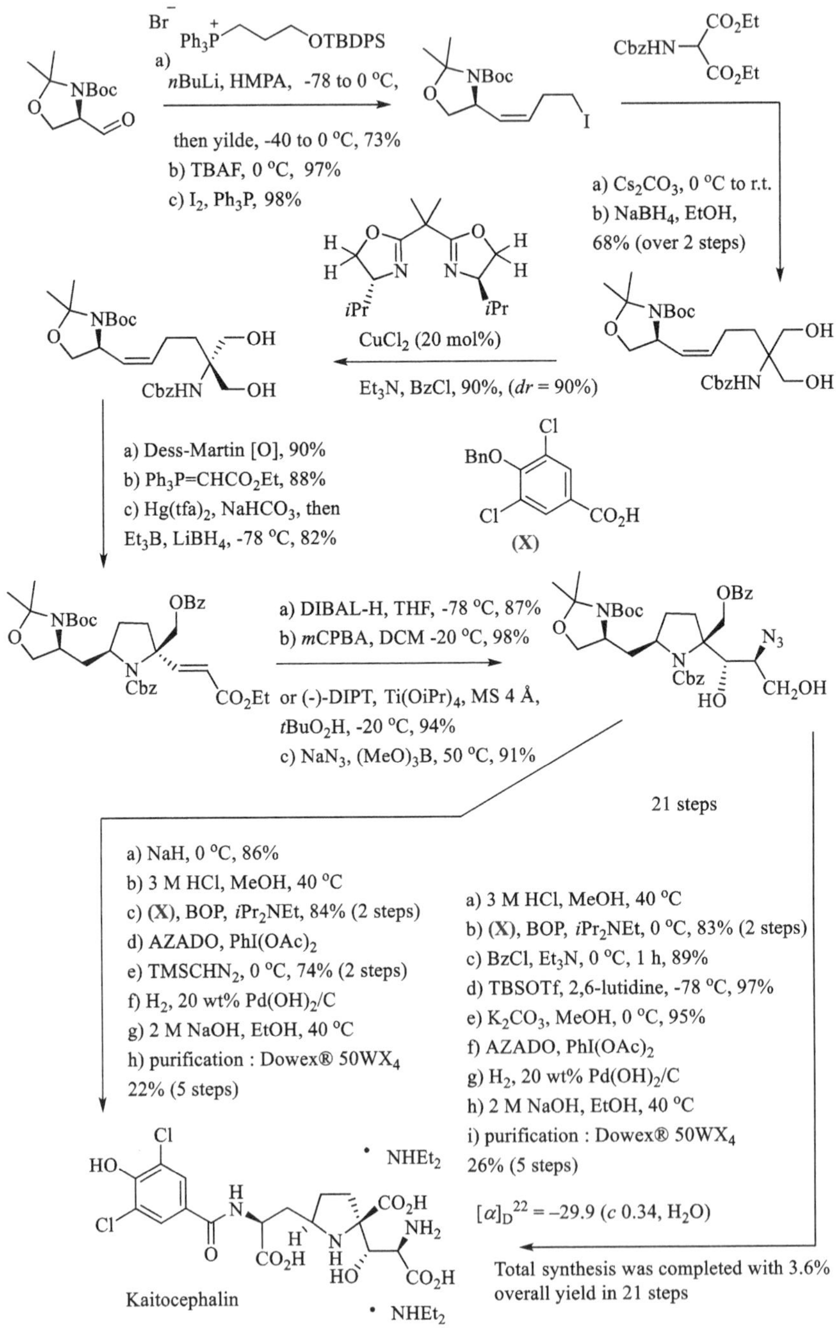


**SCHEME 69** Total synthesis.[15]

# References


1. Shin-Ya, K.; Kim, J.-S.; Furihata, K.; Hayakawa, Y.; Seto, H. Structure of Kaitocephalin, a Novel Glutamate Receptor Antagonist Produced by *Eupenicillium shearii*. *Tetrahedron Lett.* **1997,** *38,* 7079–7082.
2. Kobayashi, H.; Shin-ya, K.; Furihata, K.; Hayakawa, Y.; Seto, H. Absolute Configuration of a Novel Glutamate Receptor Antagonist Kaitocephalin. *Tetrahedron Lett.* **2001,** *42,* 4021–4023.
3. Okue, M.; Kobayashi, H.; Shin-ya, K.; Furihata, K.; Hayakawa, Y.; Seto, H.; Watanabe, H.; Kitahara, T. Synthesis of the Proposed Structure and Revision of Stereochemistry of Kaitocephalin. *Tetrahedron Lett.* **2002,** *43,* 857–860.
4. Shin-Ya, K. Novel Antitumor and Neuroprotective Substances Discovered by Characteristic Screenings Based on Specific Molecular Targets. *Biosci. Biotechnol. Biochem.* **2005,** *69,* 867–872.
5. Limon, A.; Reyes-Ruiz, J. M.; Vaswani, R. G.; Chamberlin, A. R.; Miledi, R. Kaitocephalin Antagonism of Glutamate Receptors Expressed in Xenopus Oocytes. *ACS Chem. Neurosci.* **2010,** *1,* 175–181.
6. Watanabe, H.; Okue, M.; Kobayashi, H.; Kitahara, T. The First Synthesis of Kaitocephalin Based on the Structure Revision. *Tetrahedron Lett.* **2002,** *43,* 861–864.
7. Kawasaki, M.; Shinada, T.; Hamada, M.; Ohfune, Y. Total Synthesis of (–)-kaitocephalin. *Org. Lett.* **2005,** *7,* 4165–4167.
8. Vaswani, R. G.; Chamberlin, A. R. Stereocontrolled Total Synthesis of (–)-kaitocephalin. *J. Org. Chem.* **2008,** *73,* 1661–1681.
9. Hamada, M.; Shinada, T.; Ohfune, Y. Efficient Total Synthesis of (–)-kaitocephalin. *Org. Lett.* **2009,** *11,* 4664–4667.
10. Yu, S.; Zhu, S.; Pan, X.; Yang, J.; Ma, D. Reinvestigation on Total Synthesis of Kaitocephalin and its Isomers. *Tetrahedron* **2011,** *67,* 1673–1680.
11. Takahashi, K.; Yamaguchi, D.; Ishihara, J.; Hatakeyama, S. Total Synthesis of (–)-Kaitocephalin Based on a Rh-Catalyzed C–H Amination. *Org. Lett.* **2012,** *14,* 1644–1647.
12. Lee, W.; Youn, J.-H.; Kang, S. H. Total Synthesis of (–)-kaitocephalin. *Chem. Commun.* **2013,** *49* (45), 5231–5233.
13. Garner, P.; Weerasinghe, L.; Van Houten, I.; Hu, J. A Concise [C+ NC+ CC] Coupling-Enabled Synthesis of Kaitocephalin. *Chem. Commun.* **2014,** *50,* 4908–4910.
14. Markad, P. R.; Rohokale, R. S.; Pawar, N. J.; Dhavale, D. D. D-glucose Based Synthesis of Proline–Serine C–C Linked Central and Right Hand Core of a Kaitocephalin-A Glutamate Receptor Antagonist. *RSC Adv.* **2015,** *5,* 81162–81167.
15. Lee, W. Desymmetrization-initiated Stereocontrolled Synthesis of (–)-Kaitocephalin. *Asian J. Org. Chem.* **2019,** *8,* 1687–1697.

CHAPTER

# 6

# Lactacystin

## 6.1 Abstract

(+)-Lactacystin, pyrrolidinone-based secondary metabolite, was isolated by Omura et al.[1] from the *Streptomyces species* (Fig. 6). The natural product served as a potent and selective proteasome inhibitor, which plays an important role in cellular protein degradation, an essential process in living cells[2] to regulate the cell cycle, cell division, regulation of transcription factors, etc. However, any irregularity in the proteasome functions leads to human disease, i.e., malignancies and neurodegenerative disorders. Subsequently, the inhibition of proteasome provides a significant opportunity to treat proteasome-related diseases such as arthritis, asthma, and Alzheimer's disease.[3]

The structural framework of molecules showed the presence of γ-lactam ring linked with *N*-acetyl-L-cysteine moiety via thioester bond. The natural product bears four contiguous chiral center, among which nitrogen-bearing quaternary center is a synthetic challenge for organic chemists. Upon the elimination of cysteine residue, the natural product converted into active form *β*-lactone (also known omuralide) via lactonization. The (−)-*clasto*-lactacystin (omuralide) upon hydrolysis converted into inactive dihydroxy acid, while, when enters the cell, it acylates

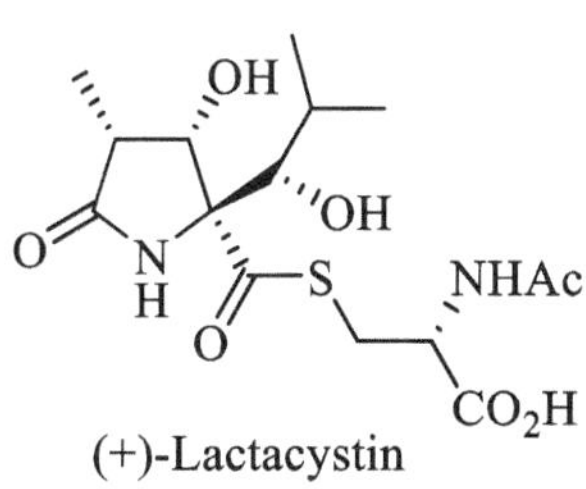


FIGURE 6 Structure of lactacystin.

α-Tertiary Amines en Route to Natural Products
https://doi.org/10.1016/B978-0-12-822262-1.00006-8

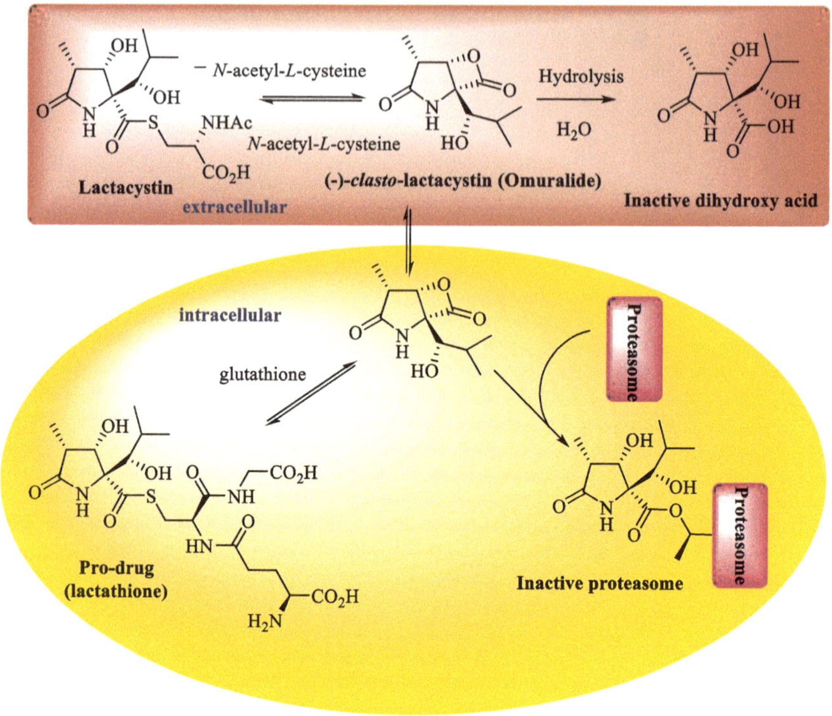


FIGURE 7 Proteasome inhibition via (−)-*clasto*-lactacystin.[6]

the proteasome to inhibit its function. Moreover, the (−)-*clasto*-lactacystin also conjugates with glutathione by thiol ester to act as reservoir for lactacystin active form, omuralide[4,5] (Fig. 7).

## 6.2 Biosynthesis of lactacystin

- The biosynthetic pathway of lactacystin was proposed by Nakagawa and coworkers.[7]
- Key precursors are isobutyric acid, L-leucine, and L-cysteine in lactacystin biosynthetic pathway.
- Key reaction involves (1) intramolecular cyclization, (2) hydroxylation of C9 carbon atom, and (3) coupling of L-cysteine **7** to afford lactacystin (Scheme 70).

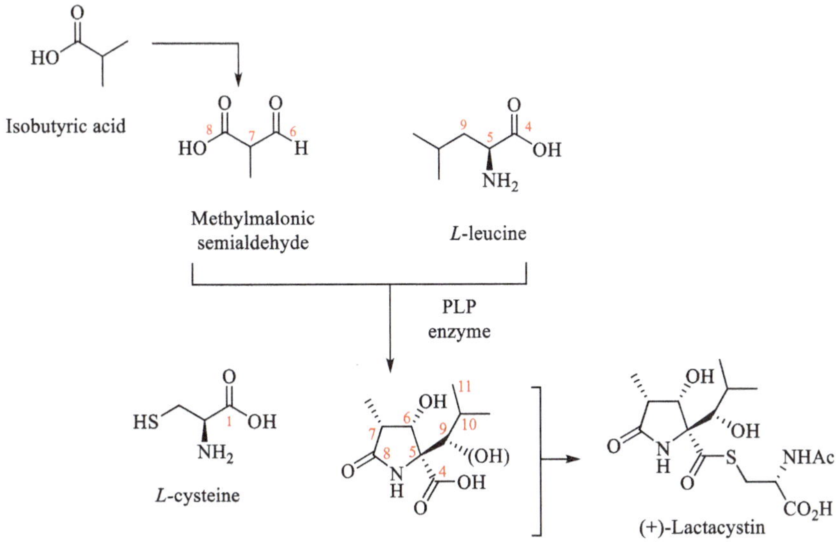


SCHEME 70 Biosynthesis of lactacystin.

## 6.3 Corey's first total synthesis of (+)-lactacystin (1992)

See (Scheme 71).

## 6.4 Corey's revised synthesis (1998)

See (Scheme 72).

## 6.5 Corey's second-generation synthesis (1998)

See (Scheme 73).

## 6.6 Corey's synthesis of α-methylomuralide (2003)

See (Scheme 74).

## 6.7 Smith-Õmura's (+)-total synthesis (1993/1996)

See (Scheme 75).

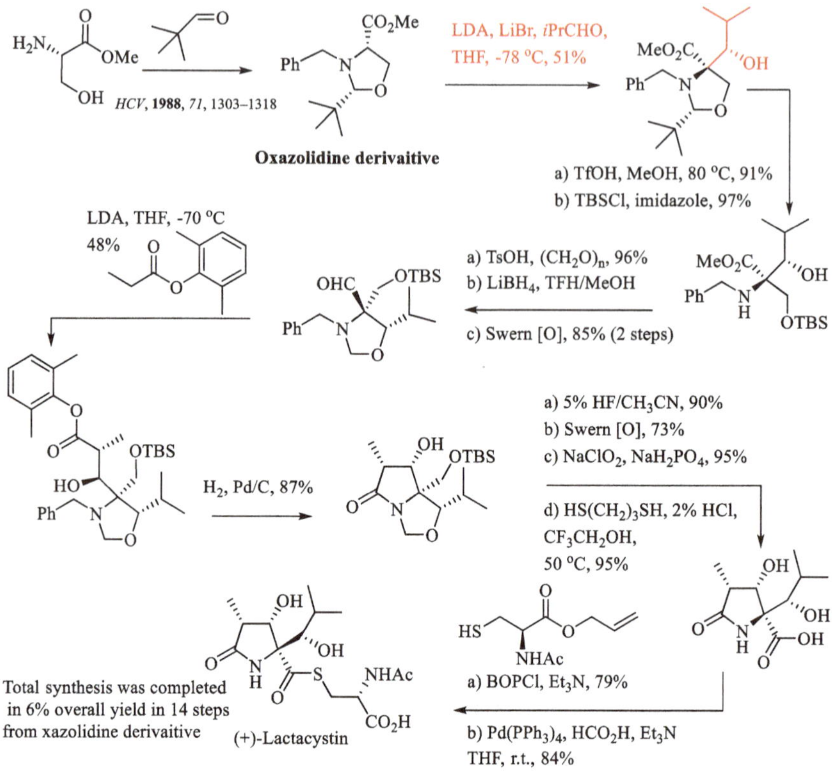


SCHEME 71 Total synthesis of (+)-lactacystin.[8]

## 6.8 Baldwin's (+)-total synthesis (1994)

See (Scheme 76).

## 6.9 Chida's (+)-total synthesis (1997)

See (Scheme 77).

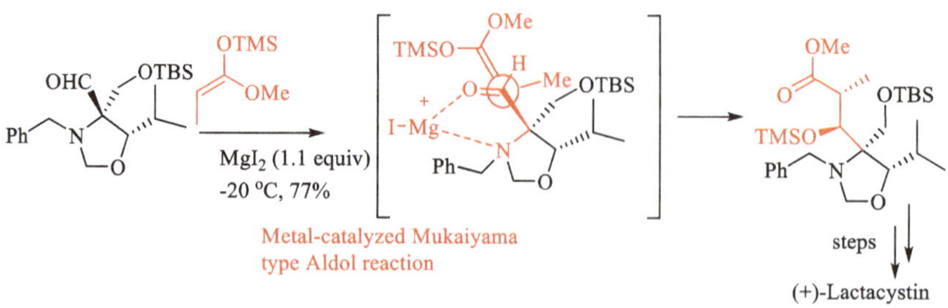


SCHEME 72 Revised synthesis of lactacystin.[9]

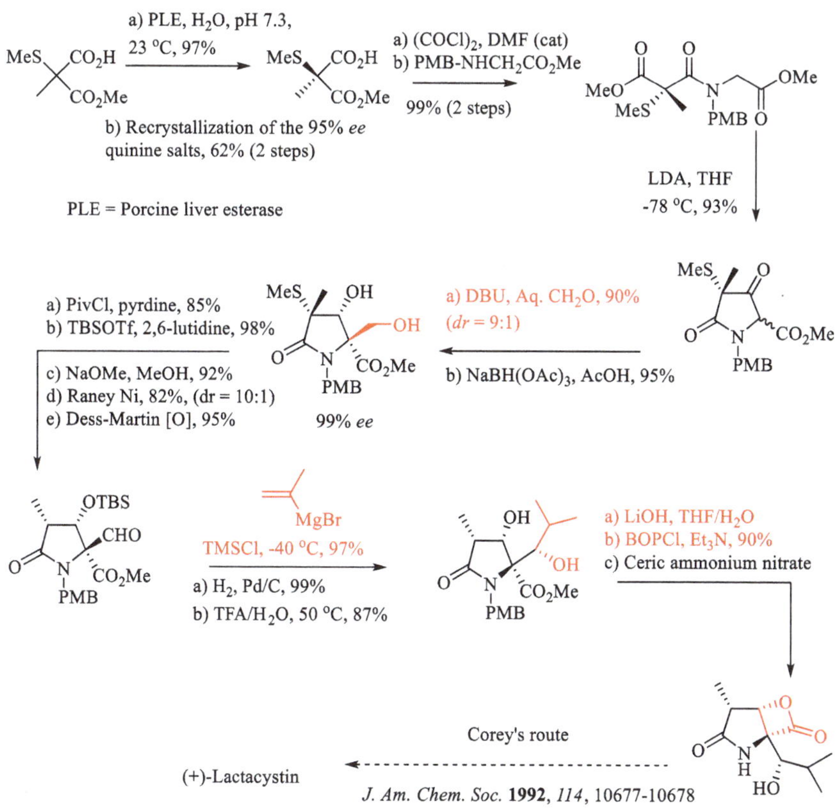


SCHEME 73 Synthesis via late-stage isopropyl group installation.[10]

## 6.10 Kang's formal synthesis (1998)

See (Scheme 78).

## 6.11 Adams *clasto*-lactacystin synthesis (1999)

See (Scheme 79).

## 6.12 Panek Total Synthesis (1999)

See (Scheme 80).

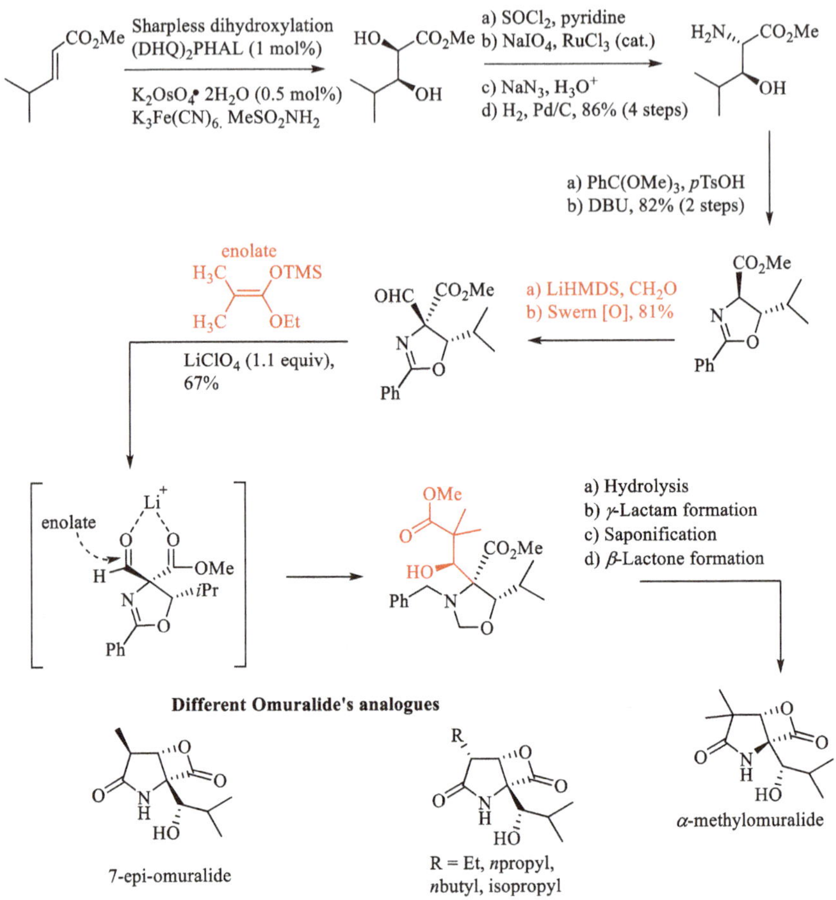


SCHEME 74 Synthesis of α-methylomuralide (2003)[11] and different analogs.[12]

## 6.13 Ohfune synthesis (2000)

See (Scheme 81).

## 6.14 Pattenden's formal synthesis (2003)

See (Scheme 82).

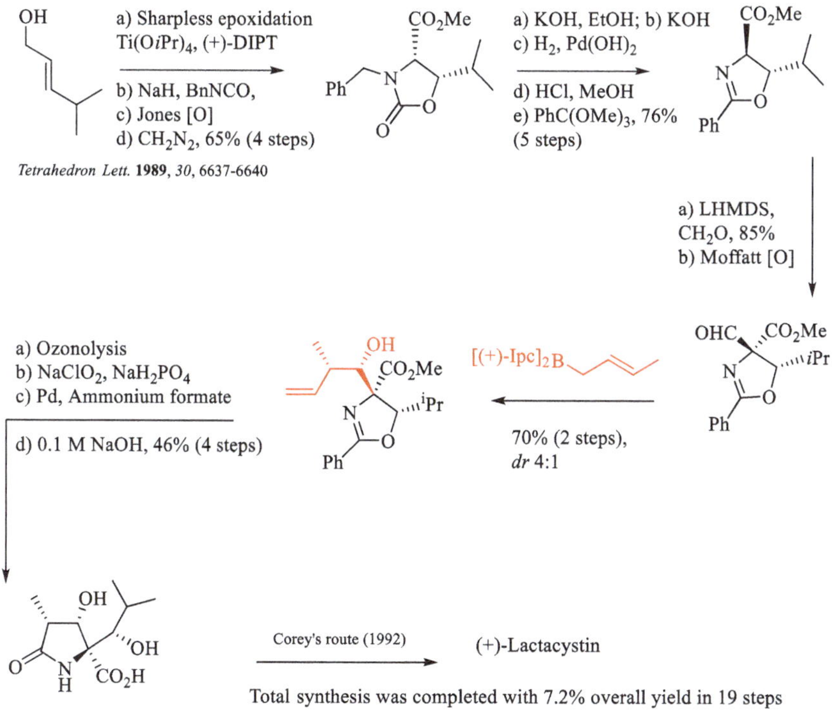


SCHEME 75 Total synthesis of lactacystin.[13]

## 6.15 Hatakeyama's total synthesis (2004)

See (Scheme 83).

## 6.16 Donohoe's racemic synthesis (2004)

See (Scheme 84).

## 6.17 Wardrop's formal synthesis (2005)

See (Scheme 85).

## 6.18 Jacobsen's total synthesis (2006)

See (Scheme 86).

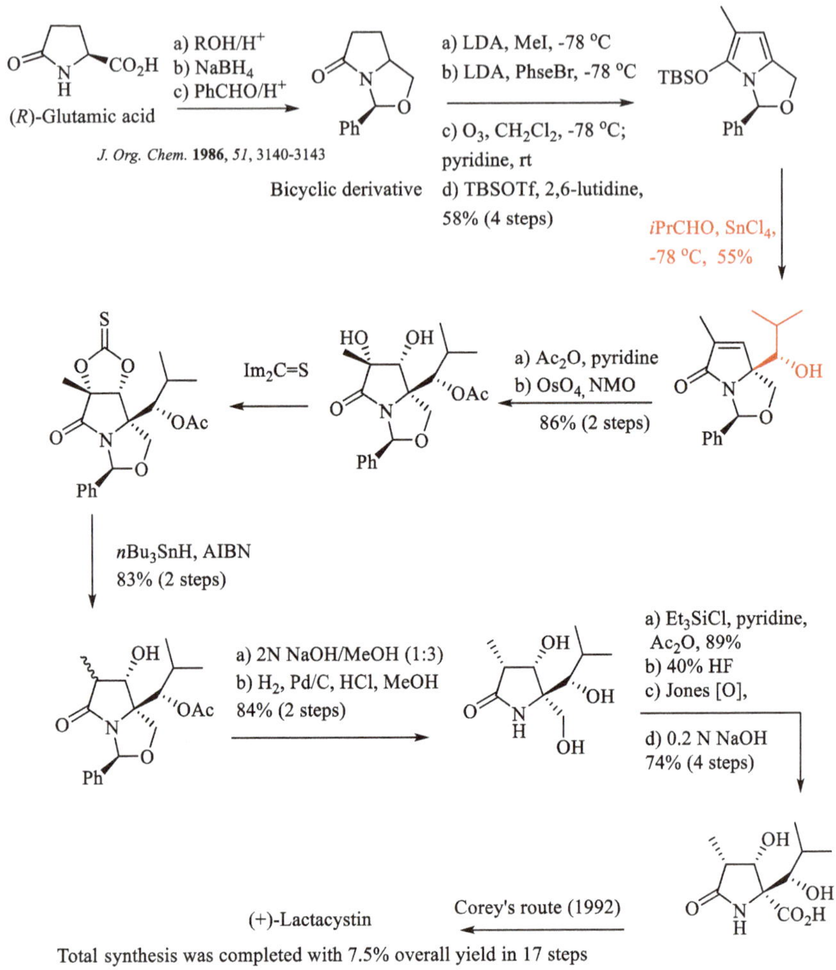


SCHEME 76 Total synthesis via SnCl4-mediated stereoselective vinylogous Mukaiyama aldol reaction.[14]

## 6.19 Shibasaki's total synthesis (2006)

See (Scheme 87).

## 6.20 Hayes's total synthesis via 1,5-CH insertion (2008)

See (Scheme 88).

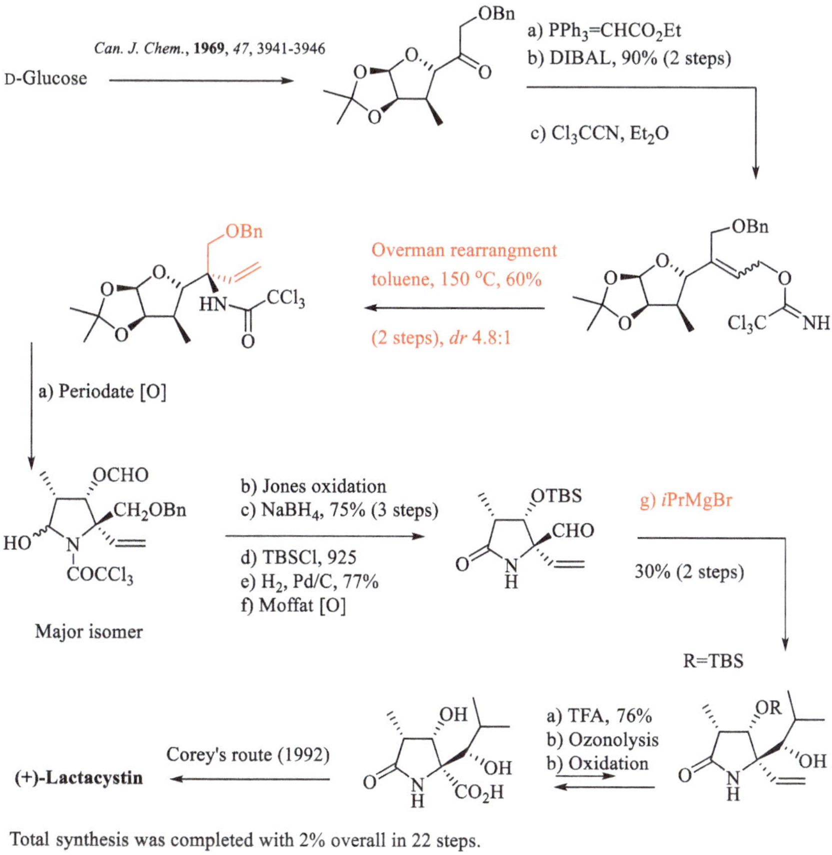


SCHEME 77 Total synthesis via Overman rearrangement.[15]

## 6.21 Hayes's formal synthesis (2010)

See (Scheme 89).

## 6.22 Silverman's total synthesis (2011)

See (Scheme 90).

## 6.23 Inoue's total synthesis (2015)

See (Scheme 91).

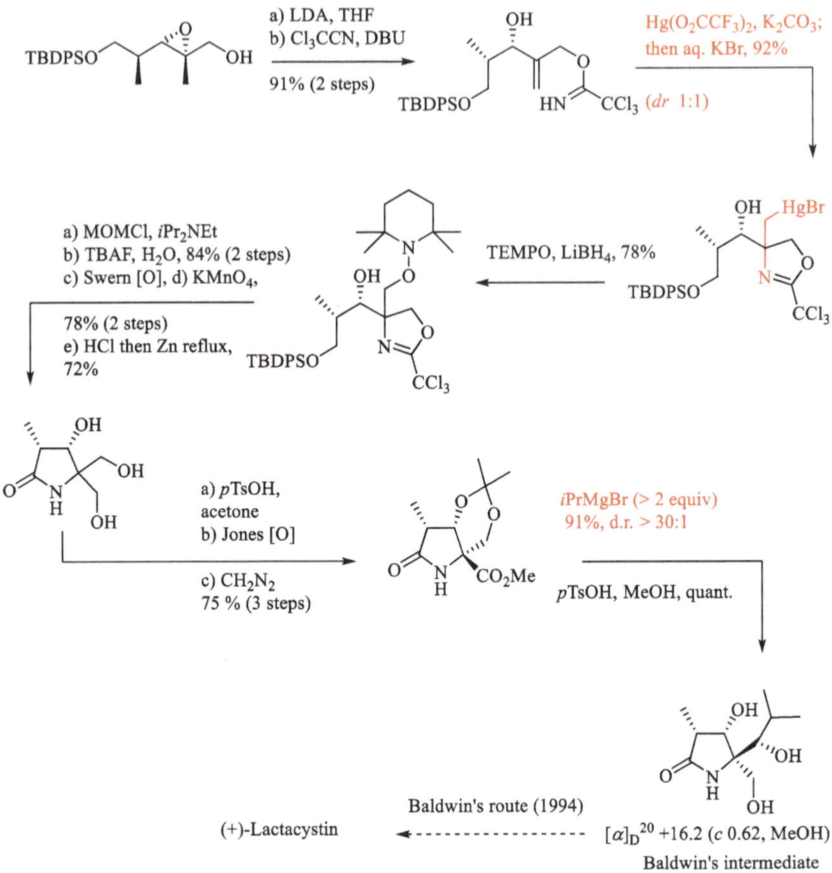


SCHEME 78 Formal synthesis via intramolecular mercurioamidation.[16]

## 6.24 Chandrasekhar's formal synthesis (2019)

See (Scheme 92).

## 6.25 Page's formal synthesis (2019)

See (Scheme 93).

## 6.26 Poisson's (−)-omuralide synthesis (2019)

See (Scheme 94).

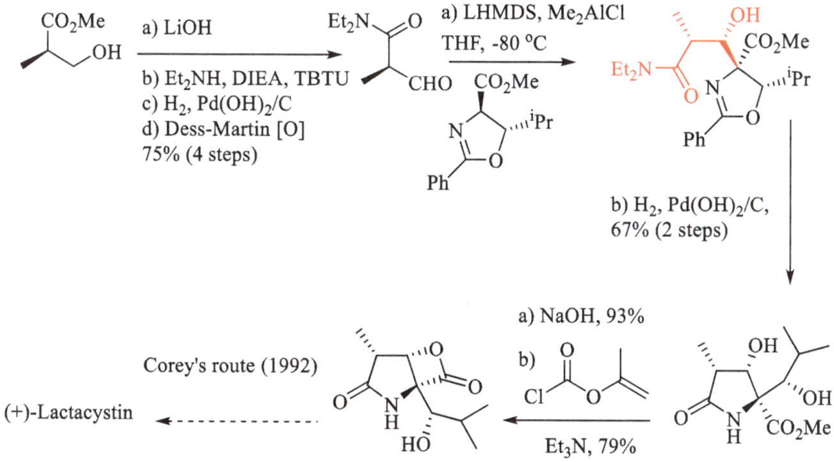


SCHEME 79 Clasto-lactacystin synthesis via aldol.[17]

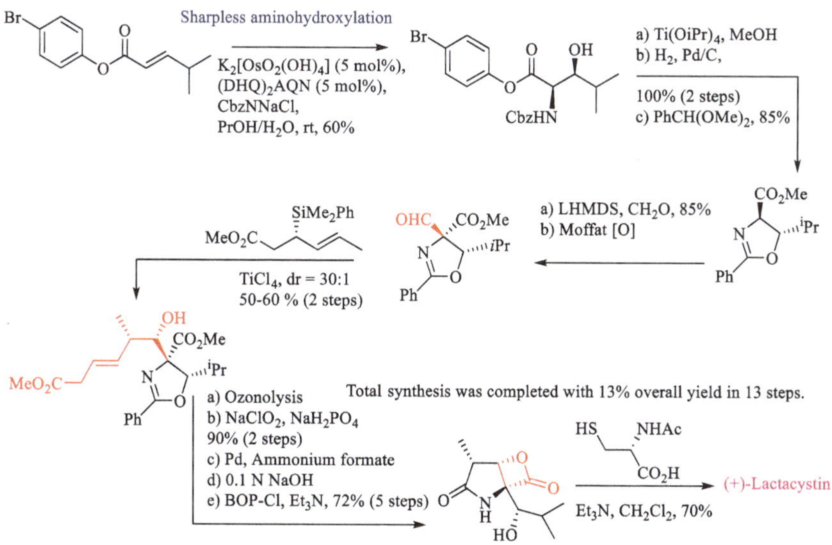


SCHEME 80 Total synthesis via asymmetric crotylation of aldehyde.[18]

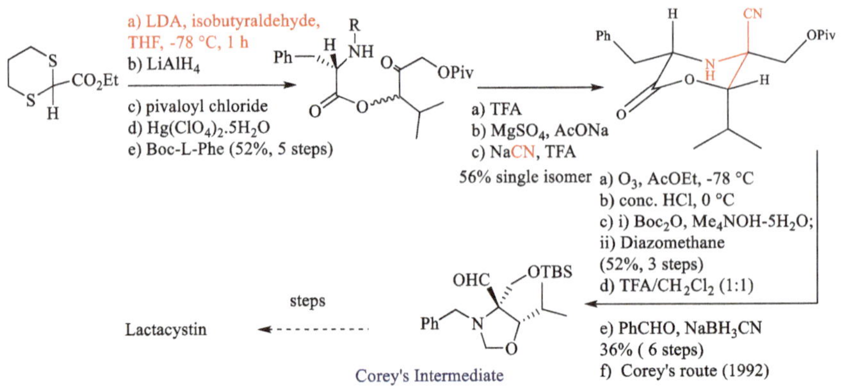


SCHEME 81 Approach toward lactacystin via Strecker reaction.[19]

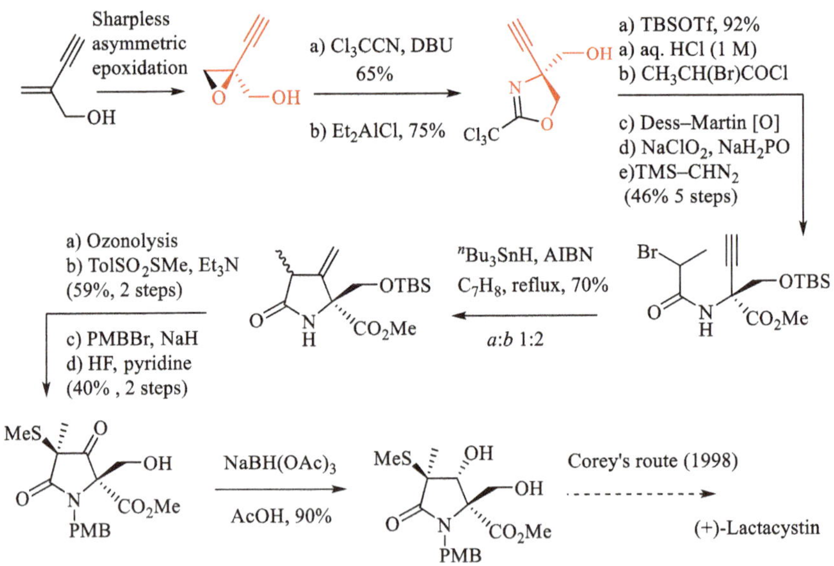


SCHEME 82 Formal synthesis via radical cyclization.[20]

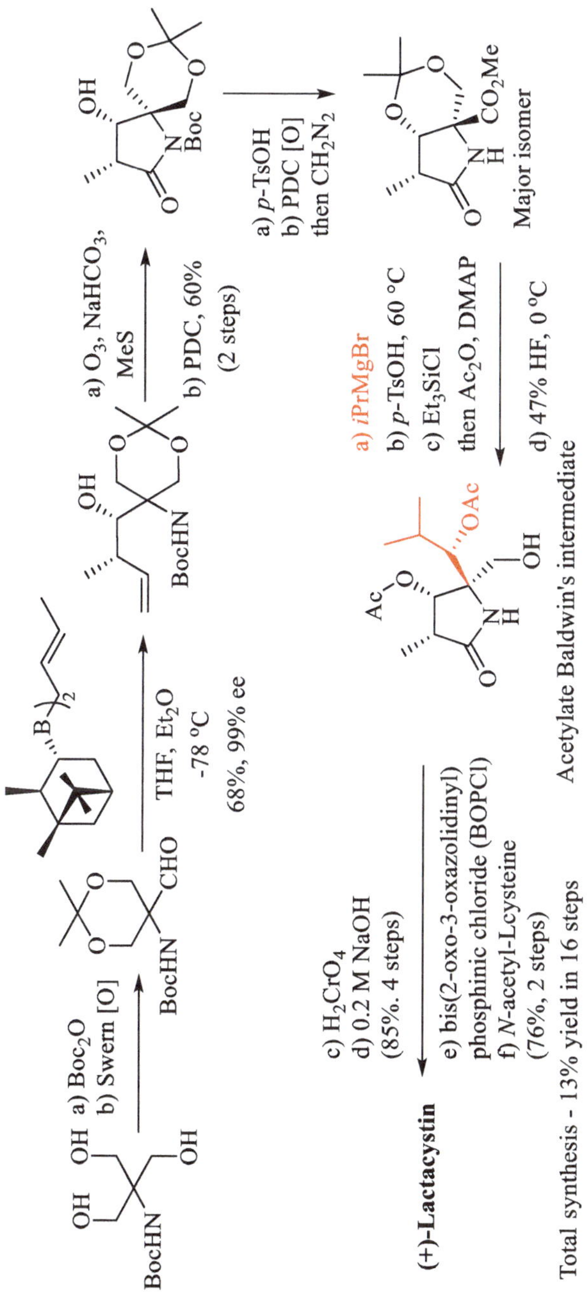


SCHEME 83 Concise total synthesis of lactacystin.[21]

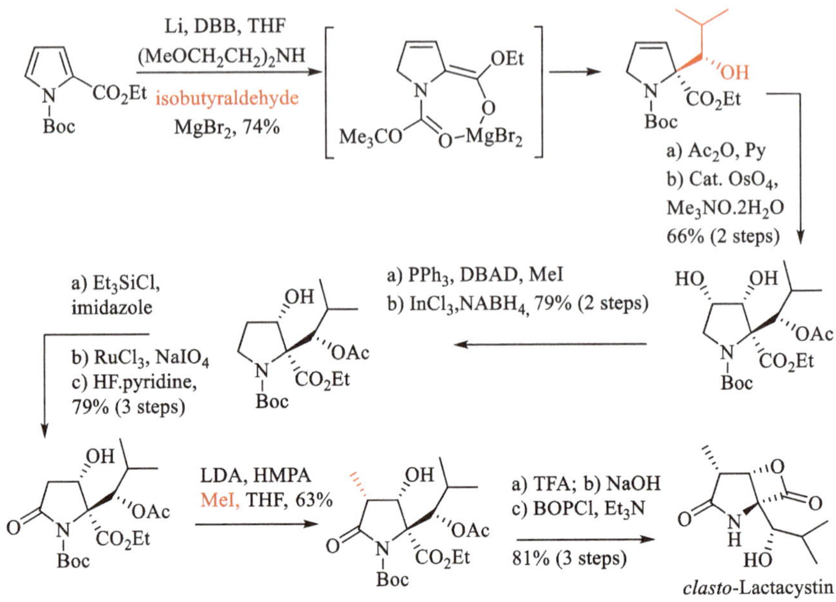


**SCHEME 84** Racemic synthesis via diastereoselective aldol reaction.[21]

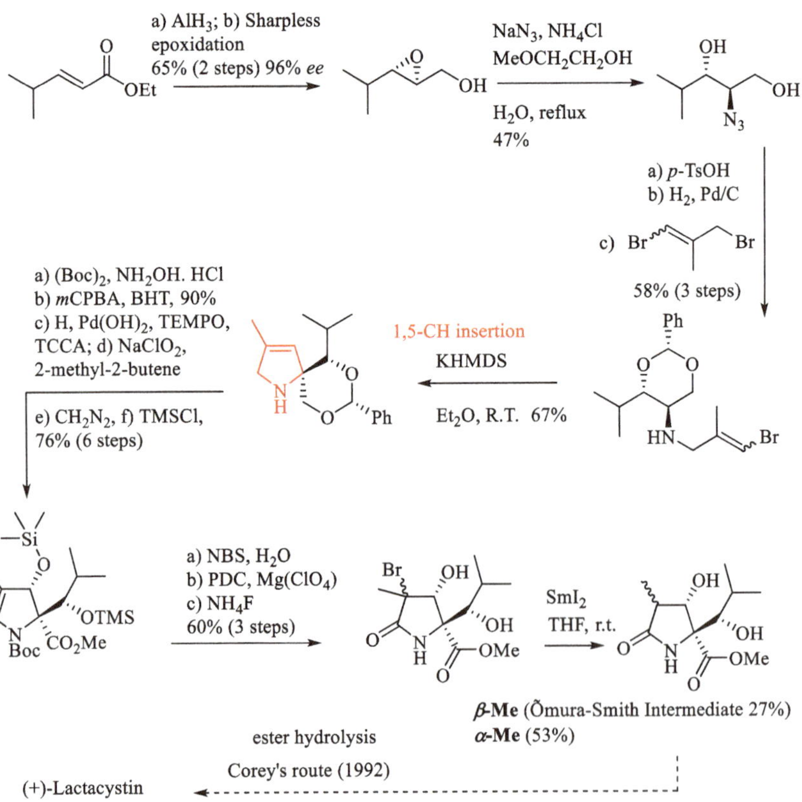


**SCHEME 85** Formal synthesis via 1,5-CH insertion.[22]

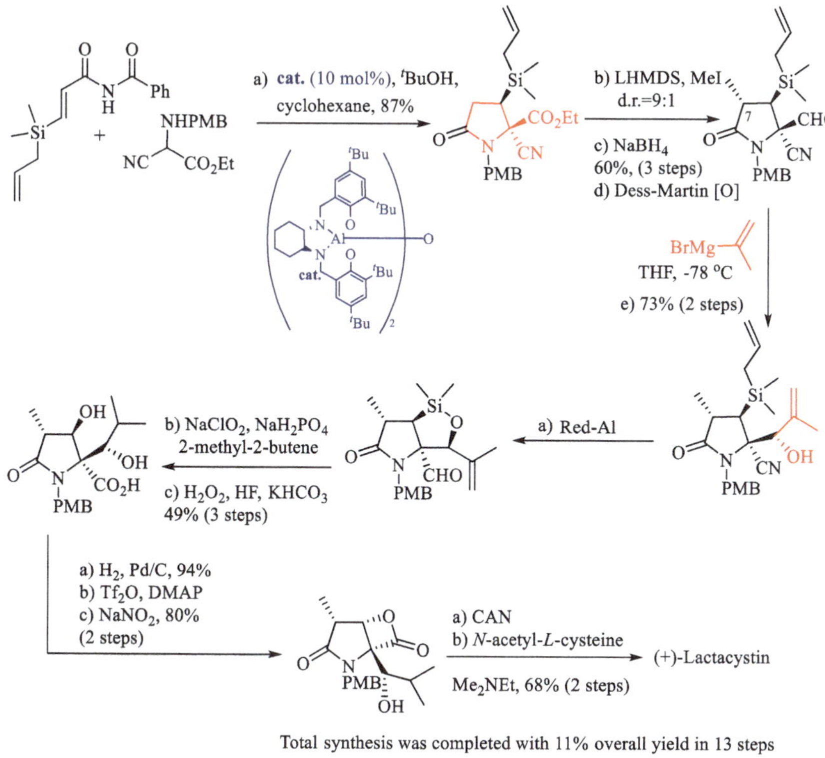


**SCHEME 86** Total synthesis via Michael addition between nitrile and $\beta$-silyl unsaturated imide.[23]

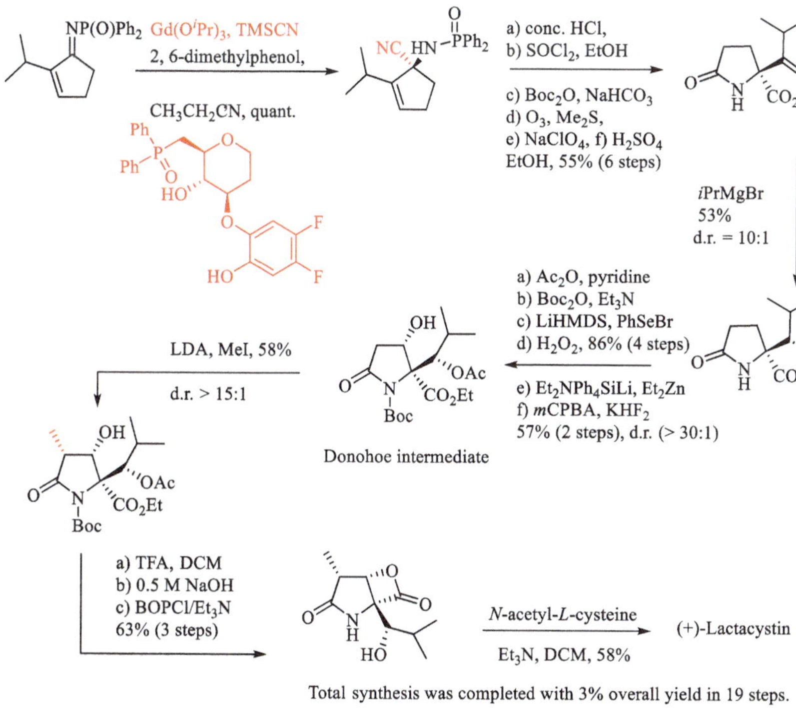


**SCHEME 87** Total synthesis via Strecker reaction.[24]

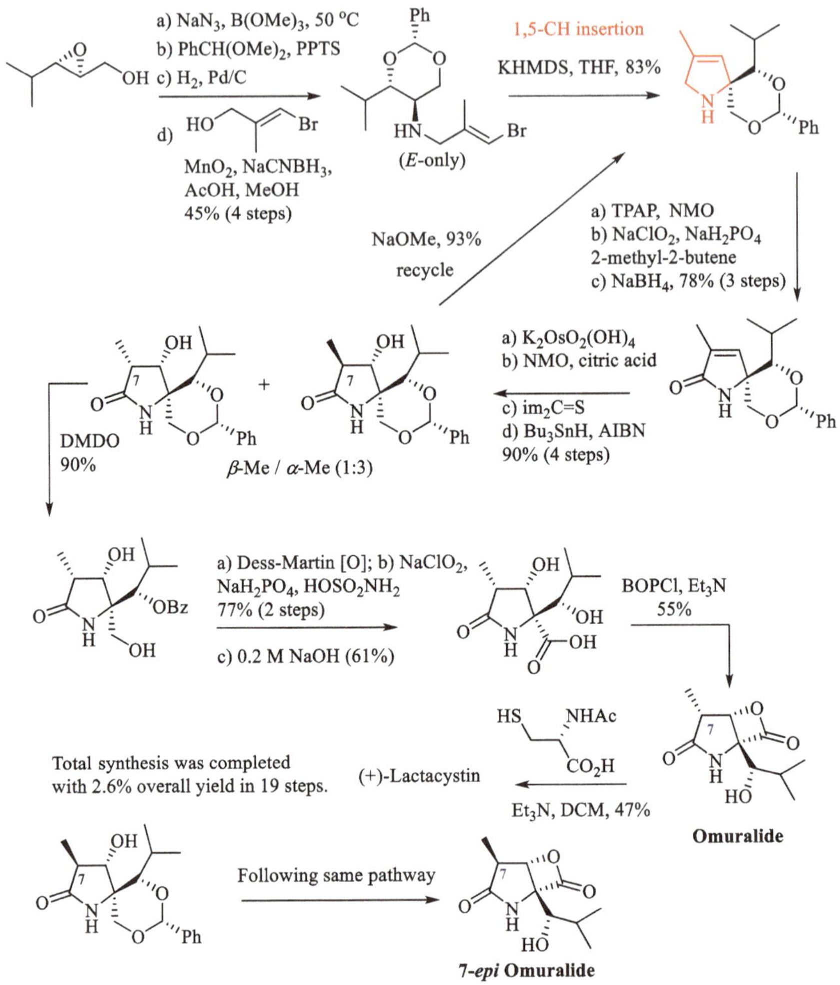


Total synthesis was completed with 2.6% overall yield in 19 steps.

SCHEME 88 Total synthesis via 1,5-CH insertion.[25]

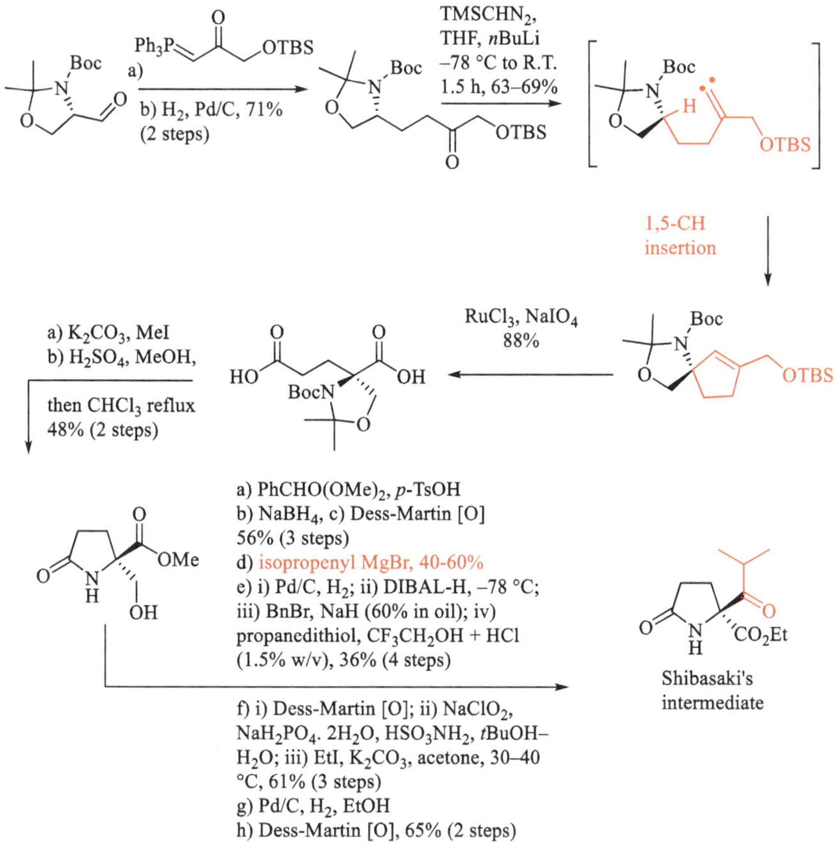


**SCHEME 89** Formal synthesis from hydroxymethyl glutamic acid (HMG) via Shibasaki's intermediate.[26]

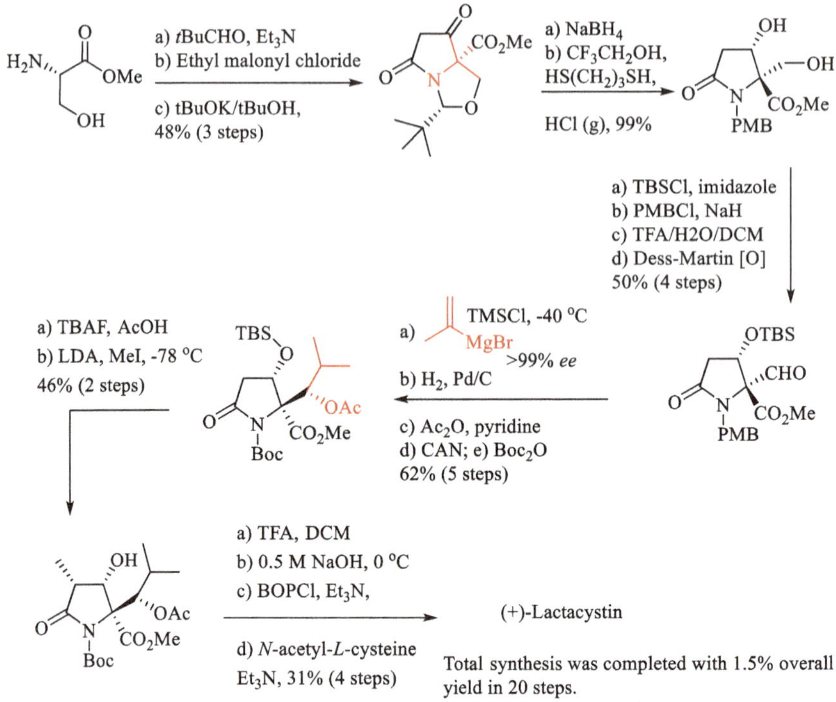


SCHEME 90 Total synthesis of lactacystin.[27]

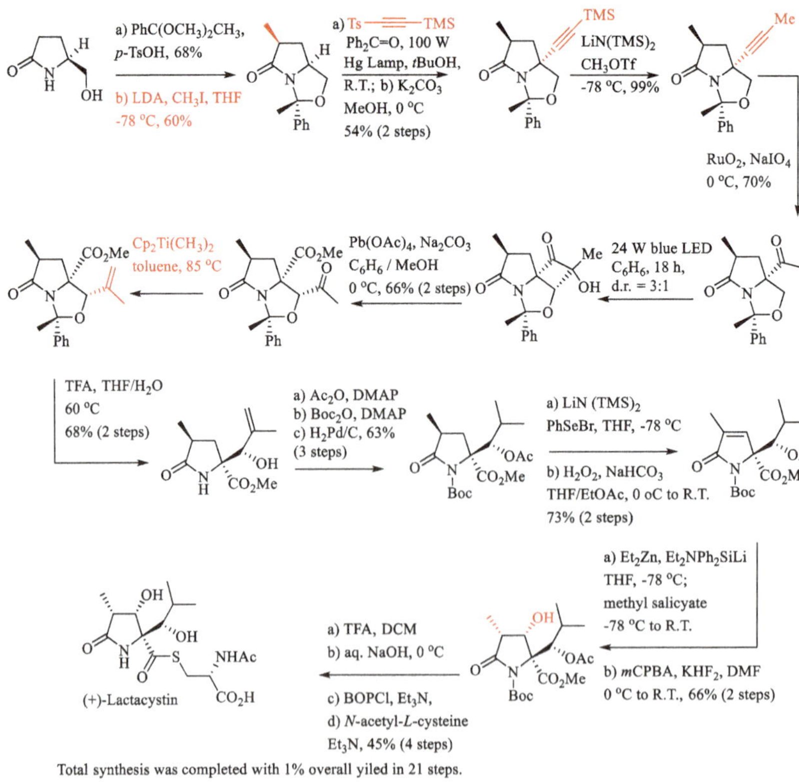


SCHEME 91 Total synthesis via C(sp$^3$)-H functionalizations.[28]

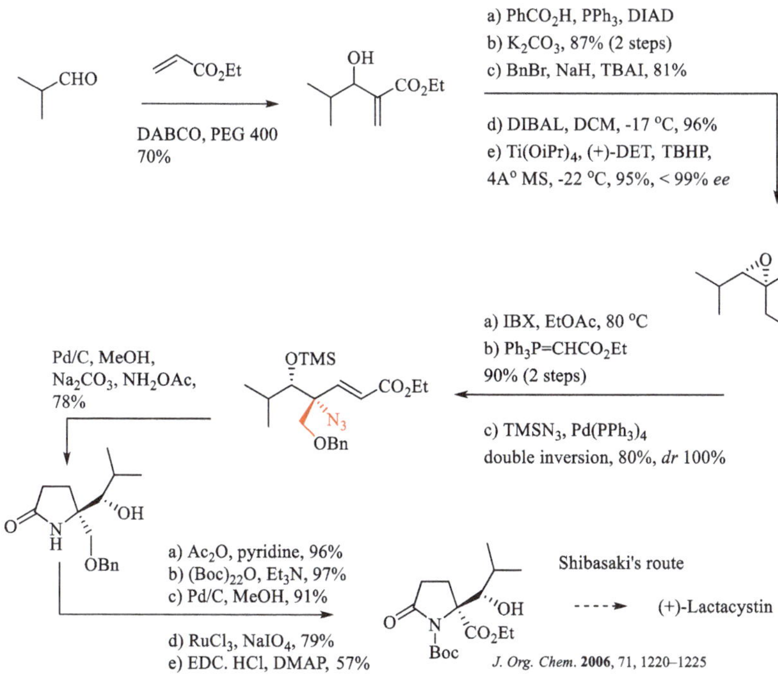


SCHEME 92 Asymmetric formal synthesis of (+)-lactacystin.[29]

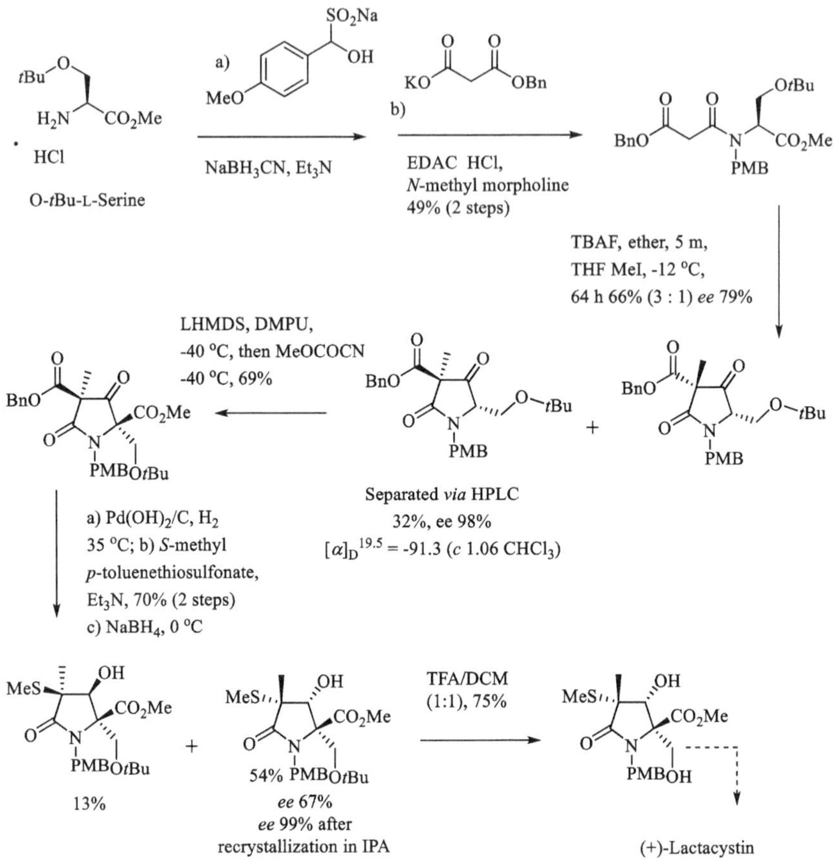


SCHEME 93 Formal synthesis of (+)-lactacystin from L-serine.[30]

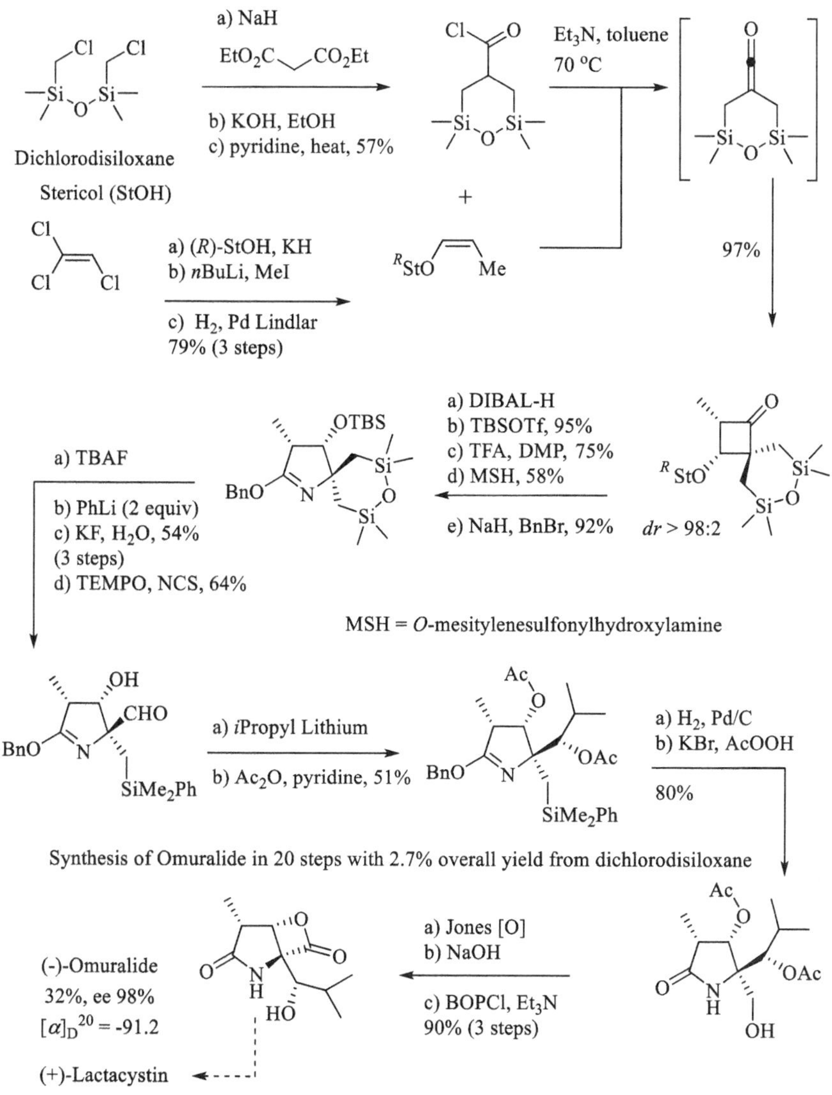


**SCHEME 94** (−)-Omuralide synthesis via asymmetric ketene [2 + 2]-cycloaddition.[31]

## References


1. Omura, S.; Fujimoto, T.; Otoguro, K.; Matsuzaki, K.; Moriguchi, R.; Tanaka, H.; Sasaki, Y. Lactacystin, a Novel Microbial Metabolite, Induces Neuritogenesis of Neuroblastoma Cells. *J. Antibiot.* **1991,** *44,* 113–116.
2. Ciechanover, A. Intracellular Protein Degradation: from a Vague Idea, through the Lysosome and the Ubiquitin–Proteasome System, and onto Human Diseases and Drug Targeting (Nobel Lecture). *Angew. Chem. Int. Ed.* **2005,** *44,* 5944–5967.
3. (a) Hefti, F.; Weiner, W. J. Nerve Growth Factor and Alzheimer's Disease. *Ann. Neurol.* **1986,** *20,* 275–281.
   (b) Doherty, P.; Dickson, J. G.; Flanigan, T. P.; Walsh, F. S. Ganglioside GM1 Does Not Initiate, but Enhances Neurite Regeneration of Nerve Growth Factor-dependent Sensory Neurones. *J. Neurochem.* **1985,** *44,* 1259–1265.
4. Dick, L. R.; Cruikshank, A. A.; Grenier, L.; Melandri, F. D.; Nunes, S. L.; Stein, R. L. Mechanistic Studies on the Inactivation of the Proteasome by Lactacystin A Central Role for *Clasto*-Lactacystin *β*-Lactone. *J. Biol. Chem.* **1996,** *271,* 7273–7276.
5. Fenteany, G.; Schreiber, S. L. Lactacystin, Proteasome Function, and Cell Fate. *J. Biol. Chem.* **1998,** *273,* 8545–8548.
6. Dick, L. R.; Cruikshank, A. A.; Destree, A. T.; Grenier, L.; McCormack, T. A.; Melandri, F. D.; Nunes, S. L.; Palombella, V. J.; Parent, L. A.; Plamondon, L. Mechanistic Studies on the Inactivation of the Proteasome by Lactacystin in Cultured Cells. *J. Biol. Chem.* **1997,** *272,* 182–188.
7. Takahashi, S.; Uchida, K.; Nakagawa, A.; Miyake, Y.; Kainosho, M.; Matsuzaki, K.; Omura, S. Biosynthesis of Lactacystin. *J. Antibiot.* **1995,** *48,* 1015–1020.
8. Corey, E.; Reichard, G. A. Total Synthesis of Lactacystin. *J. Am. Chem. Soc.* **1992,** *114* (26), 10677–10678.
9. Corey, E.; Li, W.; Reichard, G. A. A New Magnesium-Catalyzed Doubly Diastereoselective Anti-aldol Reaction Leads to a Highly Efficient Process for the Total Synthesis of Lactacystin in Quantity. *J. Am. Chem. Soc.* **1998,** *120,* 2330–2336.
10. Corey, E.; Li, W.; Nagamitsu, T. An Efficient and Concise Enantioselective Total Synthesis of Lactacystin. *Angew. Chem. Int. l Ed.* **1998,** *37,* 1676–1679.
11. Saravanan, P.; Corey, E. A Short, Stereocontrolled, and Practical Synthesis of α-methylomuralide, a Potent Inhibitor of Proteasome Function. *J. Org. Chem.* **2003,** *68* (7), 2760–2764.
12. Corey, E.; Li, W.-D. Z.; Nagamitsu, T.; Fenteany, G. The Structural Requirements for Inhibition of Proteasome Function by the Lactacystin-Derived β-lactone and Synthetic Analogs. *Tetrahedron* **1999,** *55* (11), 3305–3316.
13. Sunazuka, T.; Nagamitsu, T.; Matsuzaki, K.; Tanaka, H.; Omura, S.; Smith, A. B., III Total Synthesis of (+)-lactacystin, the First Non-protein Neurotrophic Factor. *J. Am. Chem. Soc.* **1993,** *115,* 5302-5302. Nagamitsu, T.; Sunazuka, T.; Tanaka, H.; Ōmura, S.; Sprengeler, P. A.; Smith, A. B., Total synthesis of (+)-lactacystin. J. Am. Chem. Soc. 1996, 118, 3584-3590.
14. Uno, H.; Baldwin, J. E.; Russell, A. T. Total Synthesis of (+)-lactacystin from (*R*)-glutamate. *J. Am. Chem. Soc.* **1994,** *116,* 2139–2140.
15. Chida, N.; Takeoka, J.; Ando, K.; Tsutsumi, N.; Ogawa, S. Stereoselective Total Synthesis of (+)-lactacystin from D-Glucose. *Tetrahedron* **1997,** *53* (48), 16287–16298.
16. Kang, S. An Enantiocontrolled Synthesis of a Key Intermediate to (+)-lactacystin. *Chem. Commun.* **1998,** *18,* 1929–1930.
17. Soucy, F.; Grenier, L.; Behnke, M. L.; Destree, A. T.; McCormack, T. A.; Adams, J.; Plamondon, L. A Novel and Efficient Synthesis of a Highly Active Analogue of Clasto-Lactacystin β-lactone. *J. Am. Chem. Soc.* **1999,** *121* (43), 9967–9976.

18. Masse, C. E.; Morgan, A. J.; Adams, J.; Panek, J. S. Syntheses and Biological Evaluation of (+)-Lactacystin and Analogs. *Eur. J. Org Chem.* **2000,** *2000* (14), 2513–2528.
19. Iwama, S.; Gao, W.-G.; Shinada, T.; Ohfune, Y. Asymmetric Strecker Route toward the Synthesis of the Corey Intermediate of Lactacystin. *Synlett* **2000,** *2000* (11), 1631–1633.
20. Brennan, C. J.; Pattenden, G.; Rescourio, G. Formal Synthesis of (+)-lactacystin Based on a Novel Radical Cyclisation of an α-ethynyl Substituted Serine. *Tetrahedron Lett.* **2003,** *44* (49), 8757–8760.
21. Ooi, H.; Ishibashi, N.; Iwabuchi, Y.; Ishihara, J.; Hatakeyama, S. A Concise Route to (+)-lactacystin. *J. Org. Chem.* **2004,** *69* (22), 7765–7768.
22. Wardrop, D. J.; Bowen, E. G. A Formal Synthesis of (+)-lactacystin. *Chem. Commun.* **2005,** *40,* 5106–5108.
23. Balskus, E. P.; Jacobsen, E. N. *A,* β-Unsaturated β-Silyl Imide Substrates for Catalytic, Enantioselective Conjugate Additions: A Total Synthesis of (+)-Lactacystin and the Discovery of a New Proteasome Inhibitor. *J. Am. Chem. Soc.* **2006,** *128,* 6810–6812.
24. Fukuda, N.; Sasaki, K.; Sastry, T.; Kanai, M.; Shibasaki, M. Catalytic Asymmetric Total Synthesis of (+)-lactacystin. *J. Org. Chem.* **2006,** *71,* 1220–1225.
25. Hayes, C. J.; Sherlock, A. E.; Green, M. P.; Wilson, C.; Blake, A. J.; Selby, M. D.; Prodger, J. C. Enantioselective Total Syntheses of Omuralide, 7-Epi-Omuralide, and (+)-lactacystin. *J. Org. Chem.* **2008,** *73* (6), 2041–2051.
26. Hameed, A.; Blake, A. J.; Hayes, C. J. An Enantioselective Formal Synthesis of (+)-lactacystin from Hydroxymethyl Glutamic Acid (HMG). *Synlett* **2010,** (04), 535–538.
27. Gu, W.; Silverman, R. B. Stereospecific Total Syntheses of Proteasome Inhibitors Omuralide and Lactacystin. *J. Org. Chem.* **2011,** *76,* 8287–8293.
28. Yoshioka, S.; Nagatomo, M.; Inoue, M. Application of Two Direct C($sp^3$)-H Functionalizations for Total Synthesis of (+)-Lactacystin. *Org. Lett.* **2015,** *17* (1), 90–93.
29. Sridhar, C.; Vijaykumar, B. V.; Radhika, L.; Shin, D. S.; Chandrasekhar, S. Asymmetric Formal Synthesis of (+)-Lactacystin. *Eur. J. Org Chem.* **2014,** *2014,* 6707–6712.
30. Page, P. C. B.; Goodyear, R. L.; Chan, Y.; Slawin, A. M.; Allin, S. M. Formal Synthesis of (+)-lactacystin from L-Serine. *RSC Adv.* **2019,** *9,* 30019–30032.
31. Rulliere, P.; Cannillo, A.; Grisel, J.; Cividino, P.; Carret, S.b.; Poisson, J.-F. o. Total Synthesis of Proteasome Inhibitor (−)-Omuralide through Asymmetric Ketene [2+ 2]-Cycloaddition. *Org. Lett.* **2018,** *20,* 4558–4561.

CHAPTER

# 7

# Salinosporamide A

## 7.1 Abstract

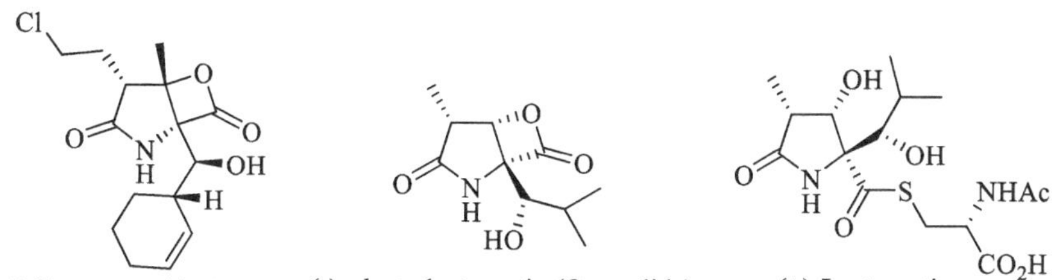

(-)-Salinosporamide A (-)-*clasto*-lactacystin (Omuralide) (+)-Lactacystin

Salinosporamide is another more potent proteasome inhibitor that was isolated by Fenical et al. from the marine microorganism, which is widely distributed in ocean sediments.[1] The natural product inhibits proteolytic activity of proteasome with $IC_{50}$ value of 1.3 nm.[2] Moreover, the salinosporamide A showed activity approximately more than 35 times compared with omuralide, an active form of lactacystin.[3] The structural scaffold of the natural product is much closely related to terrestrial microbial product omuralide, as both possess pyrrolidinone fused to a $\beta$-lactone.[4] The mechanism of action of both salinosporamide A and omuralide inhibits proteasome via the esterification of an *N*-terminal threonine residue of the 20S proteasome with the electrophilic $\beta$-lactone.[5] As the proteasome-elevated activity is involved in cancer, proteasome inhibition is a significant area of research for cancer therapy. The significant biological importance and challenge structure feature make salinosporamide an interesting synthetic target for organic chemists.


https://doi.org/10.1016/B978-0-12-822262-1.00007-X

## 7.2 Corey's first total synthesis of salinosporamide A (2004)

See (Scheme 95).

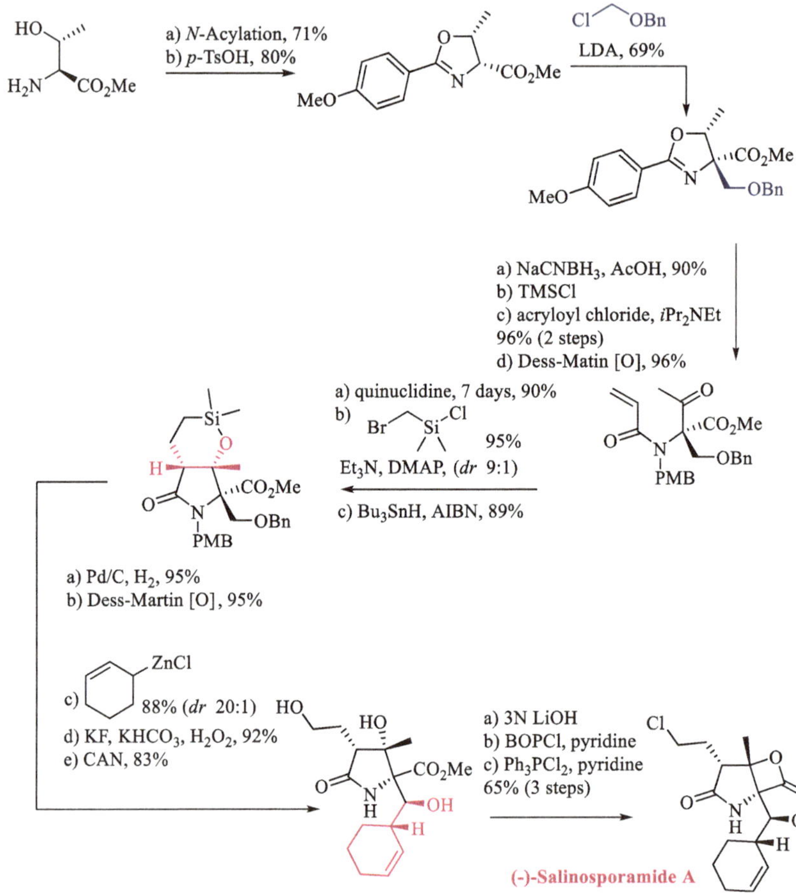


SCHEME 95 First total enantioselective synthesis of salinosporamide A.[6]

## 7.3 Second-generation improved synthesis (2005)

See (Scheme 96).

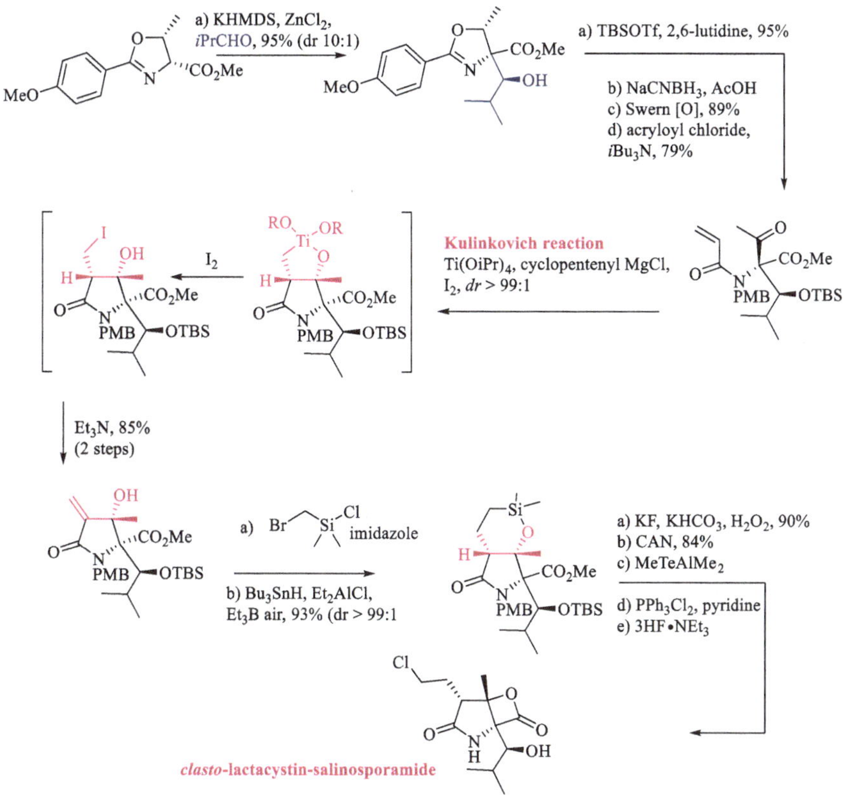


**SCHEME 96** Synthesis via Kulinkovich reaction.[7]

## 7.4 Danishefsky enantioselective synthesis (2005)

See (Scheme 97).

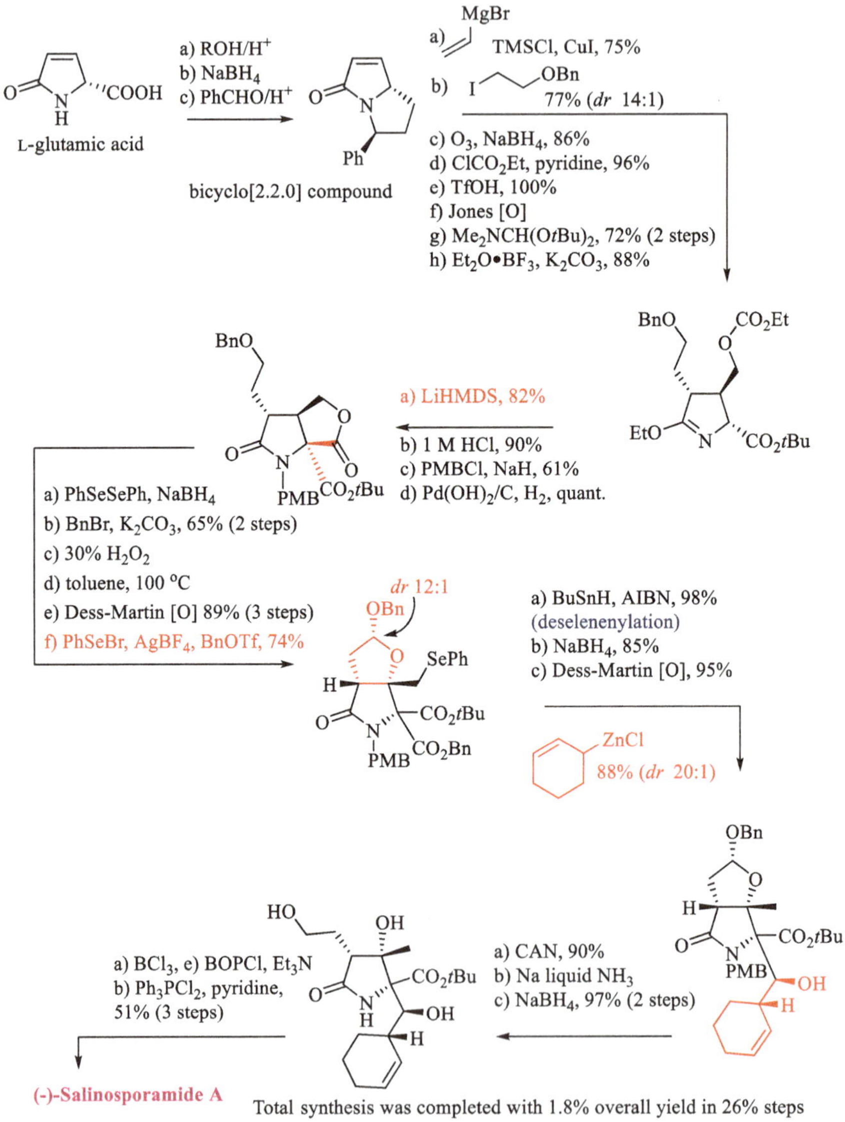


SCHEME 97 Total enantioselective synthesis.[3]

## 7.5 Pattenden racemic synthesis (2006)

See (Scheme 98).

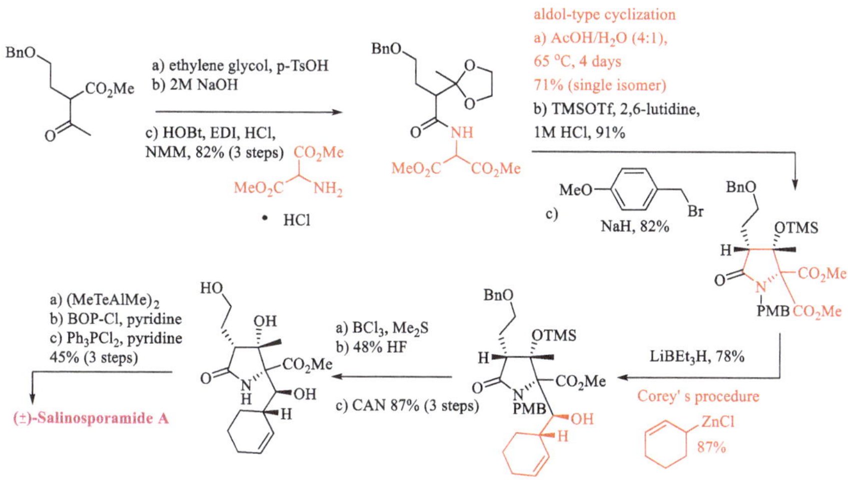


SCHEME 98 Synthesis via aldol-type cyclization.[8]

## 7.6 Lam's formal synthesis (2008)

See (Scheme 99).

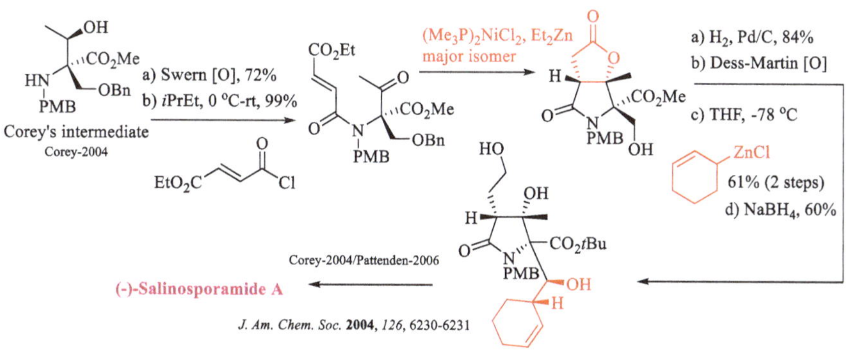


SCHEME 99 Formal synthesis via aldol cyclization–lactonization.[9]

## 7.7 Romo's asymmetric total synthesis (2011)

See (Scheme 100).

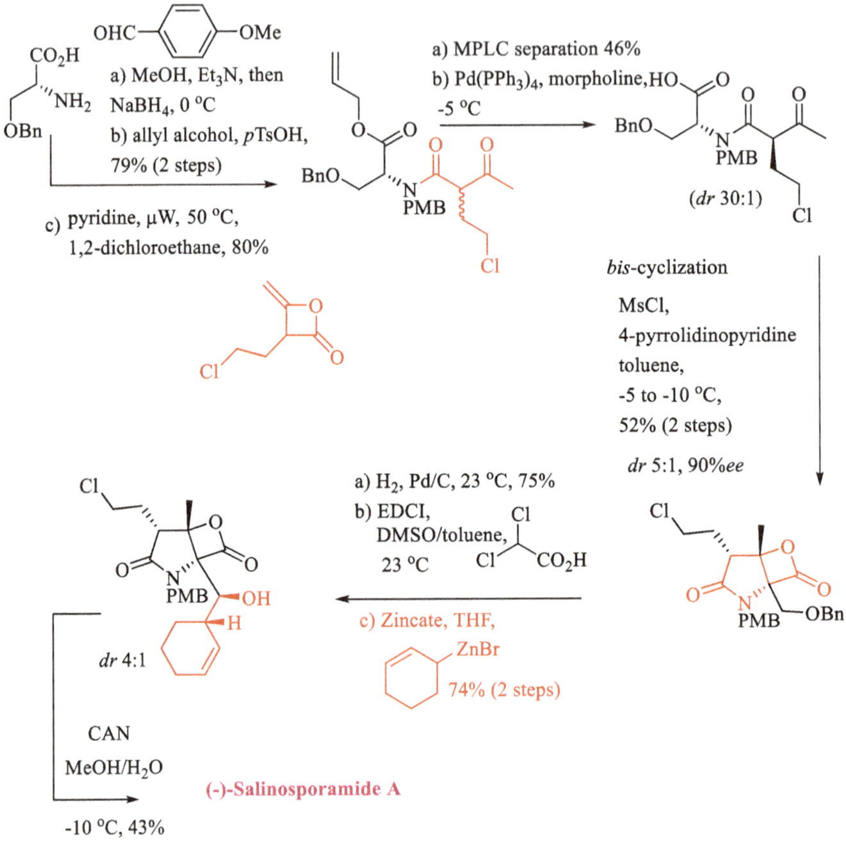


SCHEME 100 Total synthesis via bis-cyclization.[10]

# 7.8 Ling's formal synthesis (2010)

See (Scheme 101).

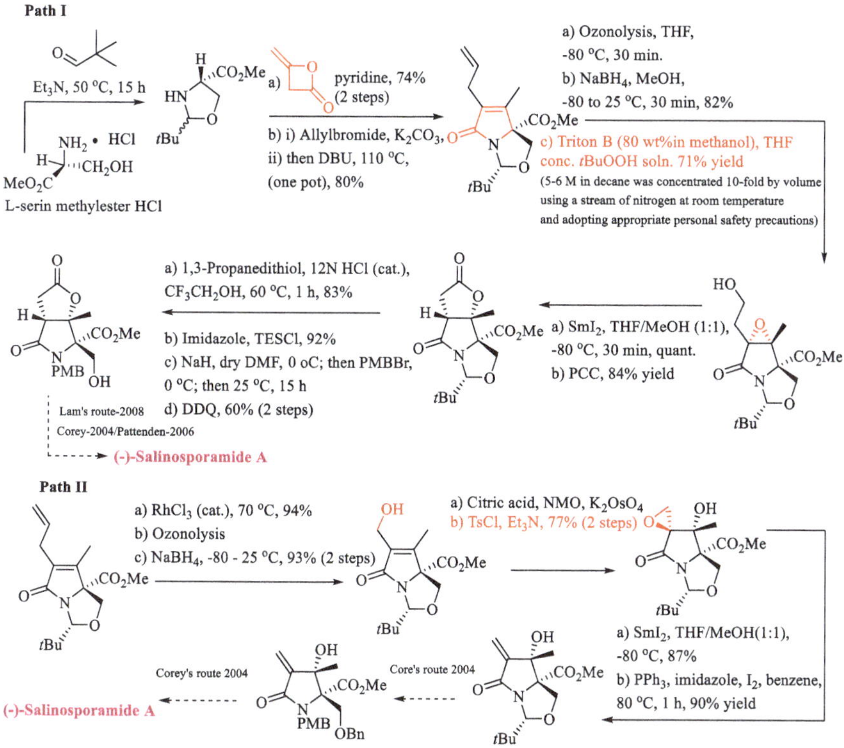


SCHEME 101 Formal synthesis via epoxidation and oxirane ring opening.[11]

## 7.9 Fukuyama's total synthesis (2011)

See (Scheme 102).

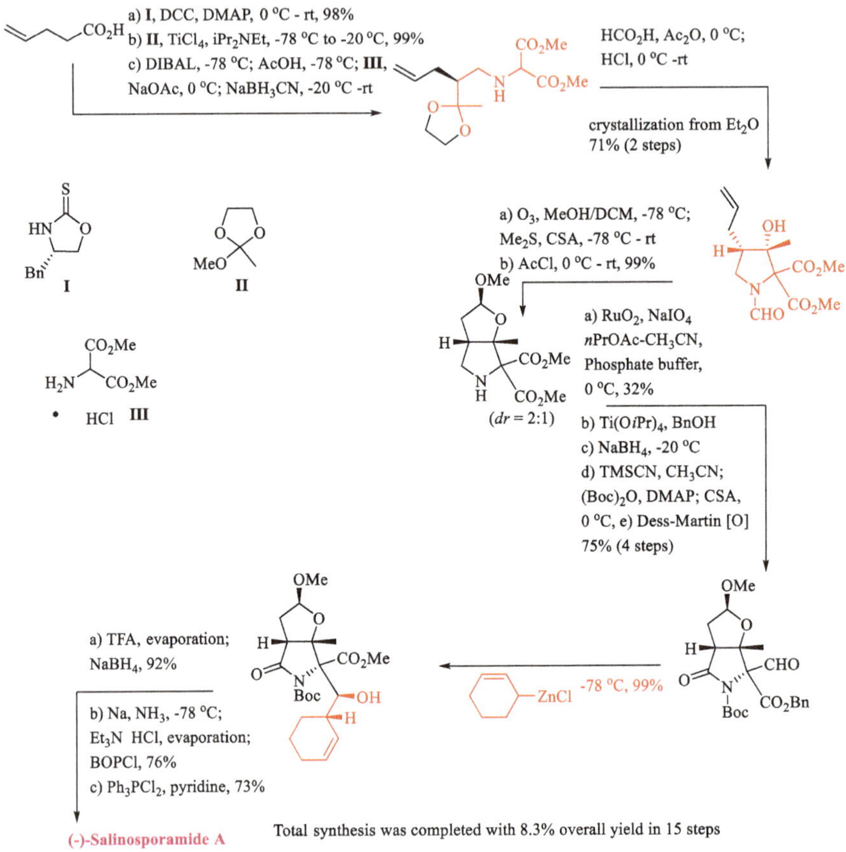


SCHEME 102 Total synthesis via alkylation with chiral auxiliary and cyclization to core pyrrolidine (a gram scale synthesis).[12]

## 7.10 Chida's total synthesis (2011)

See (Scheme 103).

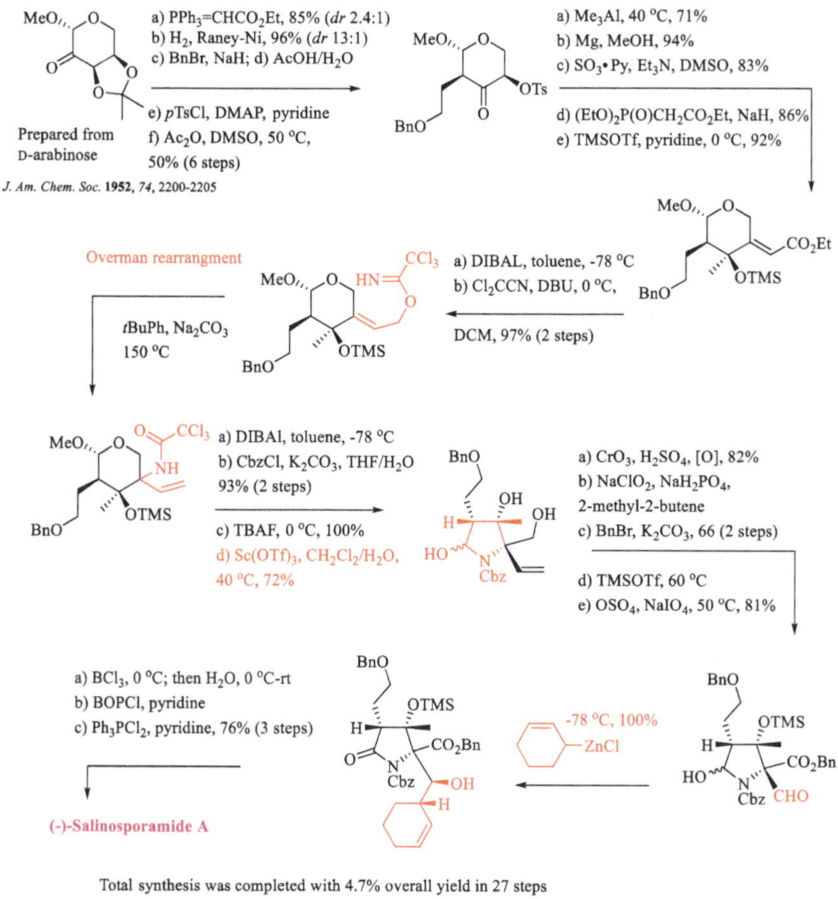


SCHEME 103 Total synthesis via Overman rearrangement.[13]

## 7.11 Lannou's approach (2012)

See (Scheme 104).

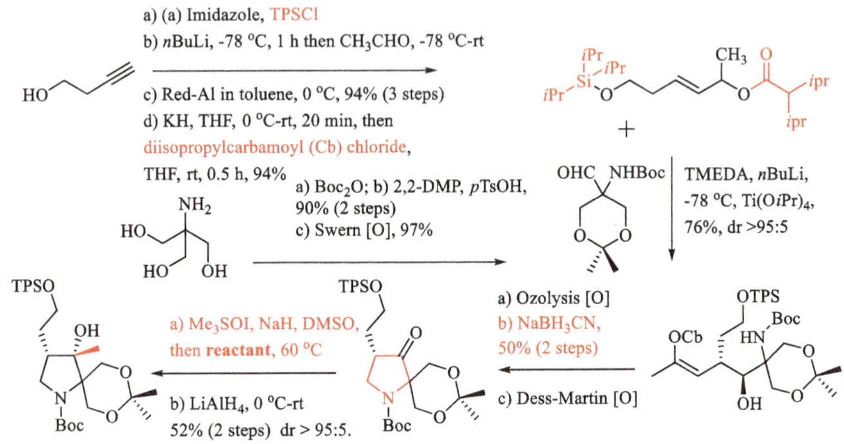


Close to Pattenden's Intermediate [*Org. Biomol. Chem.* **2006**, *4*, 2845-2846]

**SCHEME 104** Approach toward salinosporamide A.[14]

## 7.12 Burton's (−)-formal synthesis (2014)

See (Scheme 105).

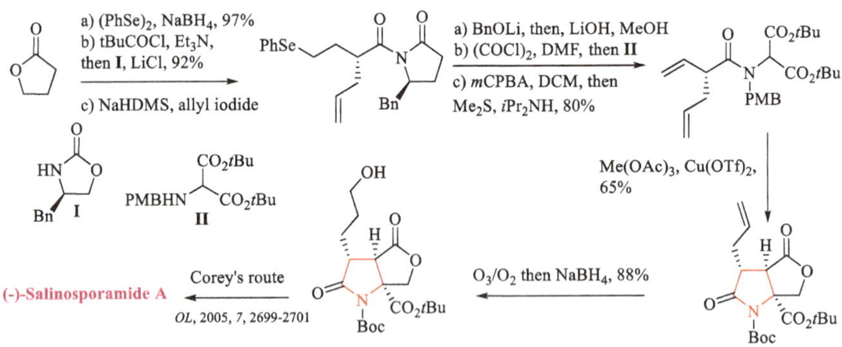


**SCHEME 105** (−)-Formal synthesis oxidative radical cyclization.[15]

## 7.13 Gonda's approach (2016)

See (Scheme 106).

- Synthesized advanced intermediate of salinosporamide A.

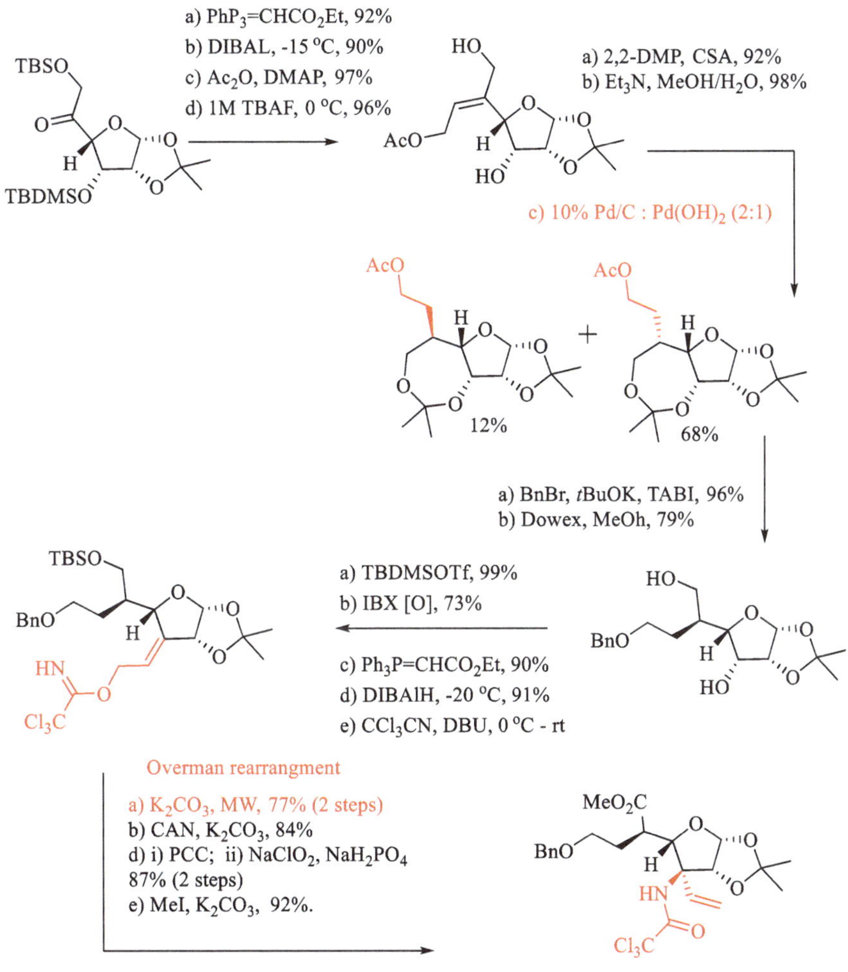


- Synthesised advance intermediate of salinosporamide A

**SCHEME 106** Approach via Overman rearrangement.[16]

## 7.14 Burton's total synthesis (2018)

See (Scheme 107).

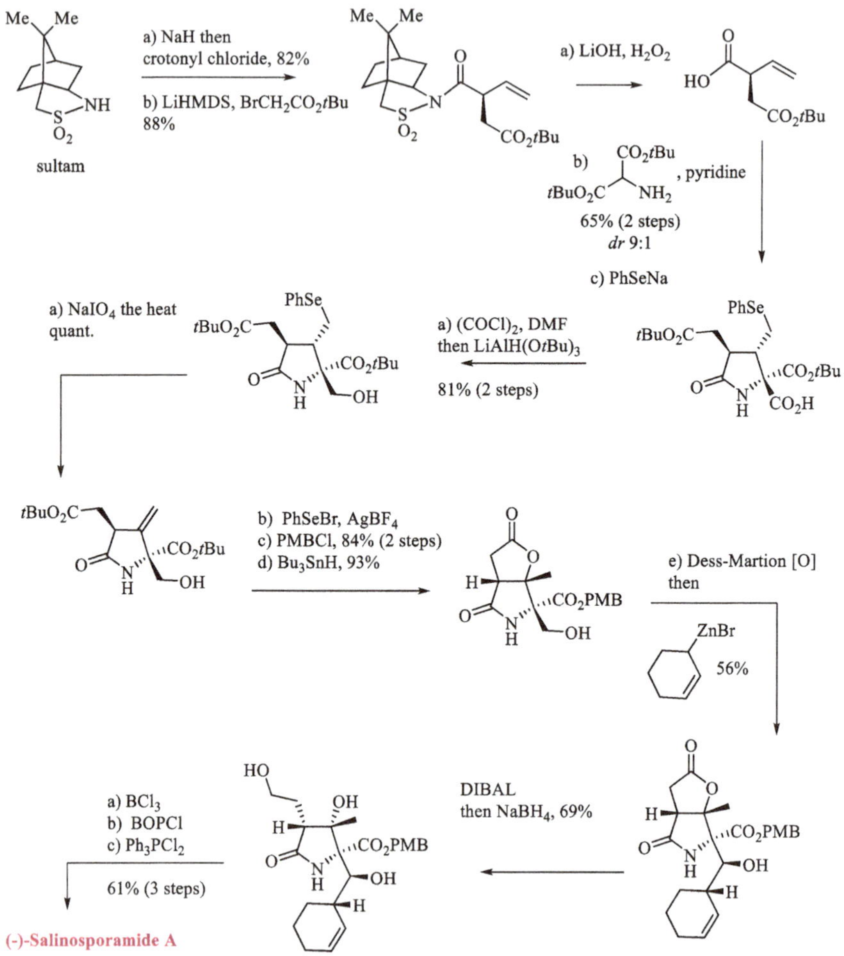


Total synthesis was completed with 5% overall yield in 16 steps (5% yield) from sultam

SCHEME 107 Total synthesis of salinosporamide A.[17]

# References


1. Feling, R. H.; Buchanan, G. O.; Mincer, T. J.; Kauffman, C. A.; Jensen, P. R.; Fenical, W.; Salinosporamide, A. A Highly Cytotoxic Proteasome Inhibitor from a Novel Microbial Source, a Marine Bacterium of the New Genus Salinospora. *Angew. Chem. Int. Ed.* **2003,** *42,* 355–357.
2. Shibasaki, M.; Kanai, M.; Fukuda, N. Total Synthesis of Lactacystin and Salinosporamide. *A.* Chem. Asian J. **2007,** *2,* 20–38.
3. Endo, A.; Danishefsky, S. J. Total Synthesis of Salinosporamide A. *J. Am. Chem. Soc.* **2005,** *127,* 8298–8299.
4. Corey, E.; Li, W.-D. Z. Total Synthesis and Biological Activity of Lactacystin, Omuralide and Analogs. *Chem. Pharm. Bull.* **1999,** *47* (1), 1–10. Reddy, L. R.; Saravanan, P.; Corey, E., A simple stereocontrolled synthesis of salinosporamide A. J. Am. Chem. Soc. 2004, 126, 6230-6231.
5. Groll, M.; Huber, R.; Potts, B. C. Crystal Structures of Salinosporamide A (NPI-0052) and B (NPI-0047) in Complex with the 20S Proteasome Reveal Important Consequences of *β*-lactone Ring Opening and a Mechanism for Irreversible Binding. *J. Am. Chem. Soc.* **2006,** *128,* 5136–5141.
6. Reddy, L. R.; Saravanan, P.; Corey, E. A Simple Stereocontrolled Synthesis of Salinosporamide A. *J. Am. Chem. Soc.* **2004,** *126,* 6230–6231.
7. Reddy, L. R.; Fournier, J.-F.; Subba Reddy, B.; Corey, E. An Efficient, Stereocontrolled Synthesis of a Potent Omuralide–salinosporin Hybrid for Selective Proteasome Inhibition. *J. Am. Chem. Soc.* **2005,** *127,* 8974–8976.
8. Mulholland, N. P.; Pattenden, G.; Walters, I. A. A Concise Total Synthesis of Salinosporamide. *A.* Org. Biomol. Chem. **2006,** *4,* 2845–2846.
9. Margalef, I. V.; Rupnicki, L.; Lam, H. W. Formal Synthesis of Salinosporamide A Using a Nickel-Catalyzed Reductive Aldol Cyclization–Lactonization as a Key Step. *Tetrahedron* **2008,** *64,* 7896–7901.
10. Nguyen, H.; Ma, G.; Gladysheva, T.; Fremgen, T.; Romo, D. Bioinspired Total Synthesis and Human Proteasome Inhibitory Activity of (−)-salinosporamide A,(−)-homosalinosporamide A, and Derivatives Obtained *via* Organonucleophile Promoted *Bis*-Cyclizations. *J. Org. Chem.* **2011,** *76,* 2–12.
11. Ling, T.; Potts, B. C.; Macherla, V. R. Concise Formal Synthesis of (−)-salinosporamide A (Marizomib) Using a Regio-And Stereoselective Epoxidation and Reductive Oxirane Ring-Opening Strategy. *J. Org. Chem.* **2010,** *75,* 3882–3885.
12. Satoh, N.; Yokoshima, S.; Fukuyama, T. Total Synthesis of (−)-Salinosporamide. *A.* Org. Lett. **2011,** *13,* 3028–3031.
13. Kaiya, Y.; Hasegawa, J.i.; Momose, T.; Sato, T.; Chida, N. Total Synthesis of (−)-Salinosporamide. *A.* Chem. Asian J. **2011,** *6,* 209–219.
14. Barbion, J.; Sorin, G.; Selkti, M.; Kellenberger, E.; Baati, R.; Santoro, S.; Himo, F.; Pancrazi, A.; Lannou, M.-I.; Ardisson, J. Stereoselective Functionalization of Pyrrolidinone Moiety towards the Synthesis of Salinosporamide A. *Tetrahedron* **2012,** *68,* 6504–6512.
15. Logan, A. W.; Sprague, S. J.; Foster, R. W.; Marx, L. B.; Garzya, V.; Hallside, M. S.; Thompson, A. L.; Burton, J. W. Diastereoselective Synthesis of Fused Lactone-Pyrrolidinones; Application to a Formal Synthesis of (−)-Salinosporamide. *A.* Org. Lett. **2014,** *16* (16), 4078–4081.
16. Takacsová, M.; Čonková, M.; Martinková, M.; Gonda, J. Convenient Approach to an Advanced Intermediate for Salinosporamide A Synthesis. *Tetrahedron: Asymmetry* **2016,** *27,* 369–376.
17. Marx, L. B.; Burton, J. W. A Total Synthesis of Salinosporamide. *A.* Chem.: Eur. J. **2018,** *24,* 6747–6754.

CHAPTER

# 8

# Manzacidins

The manzacidins A–C are a family of bromopyrrole alkaloids that are isolated by Kobayashi and coworkers from the Okinawan sponge *Hymeniacidon* species.[1] The manzacidin D without bromo atom was obtained from coralline demosponge *Astroscler willeyana*.[2] The bromopyrrole members of manzacidin family showed interesting biological activities, i.e., $\alpha$-adrenoceptor blockers, antagonists of the serotonergic receptor, and actomyosin ATPase; however, the inadequate amount of natural products from natural sources promoted their synthetic studies via novel synthetic methodologies to explore further their biological activities[3–5] (Fig. 8).

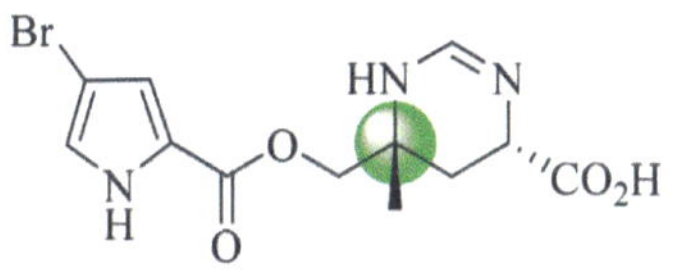

**Manzacidine A**

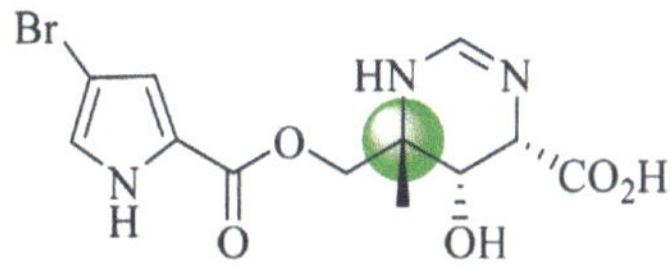

**Manzacidine B**

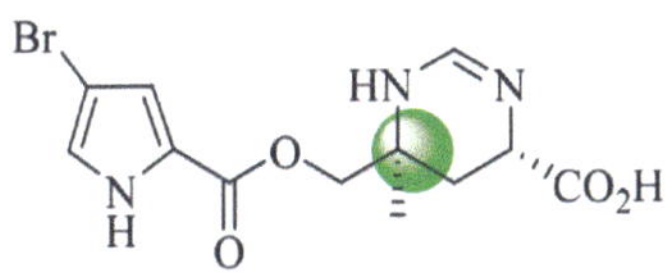

**Manzacidine C**

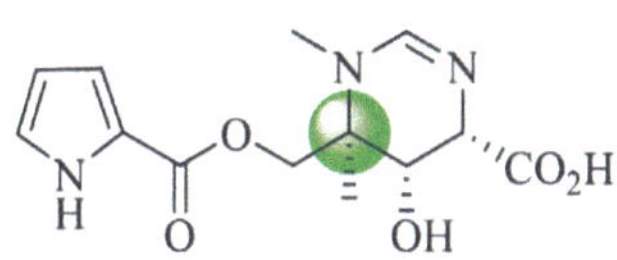

**Manzacidine D**

FIGURE 8 Structures of manzacidins.


https://doi.org/10.1016/B978-0-12-822262-1.00008-1

## 8.1 Ohfune's total synthesis of manzacidin A and C (2000)

See (Scheme 108).

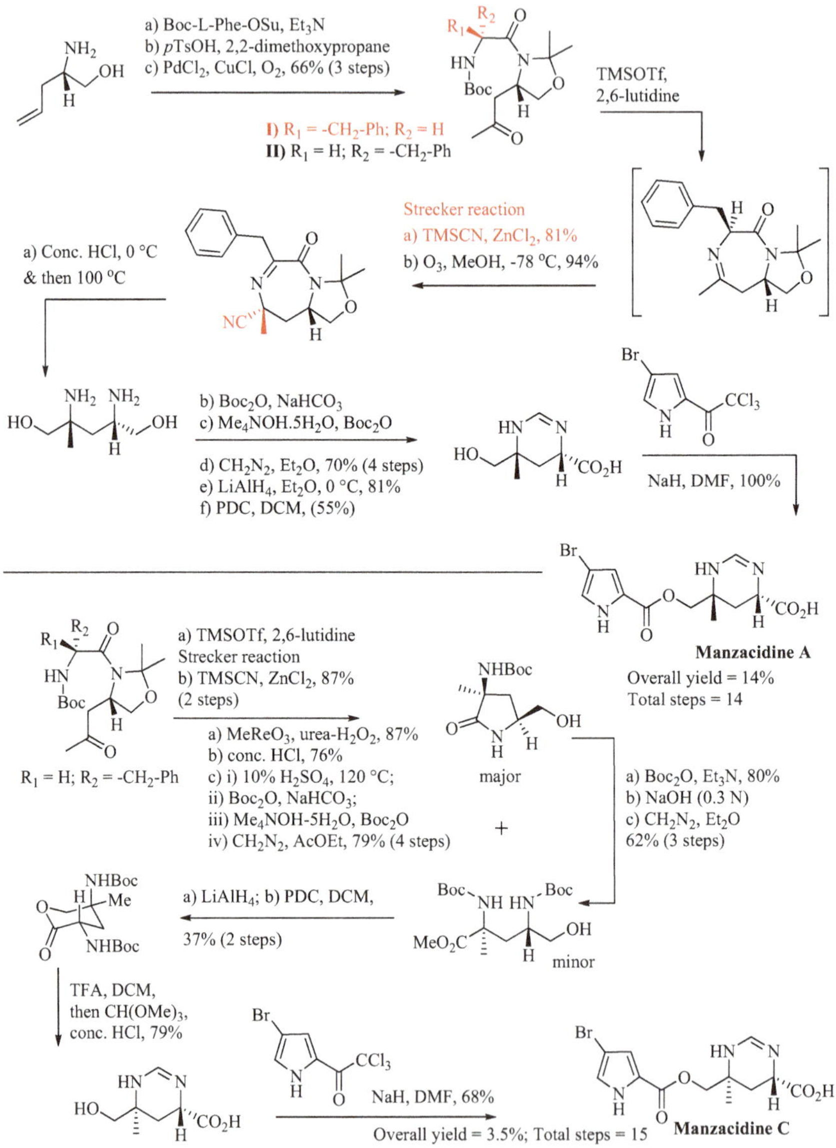


**SCHEME 108** Total synthesis of manzacidin A and C via Strecker reaction.[6]

# 8.2 Du Bois' enantioselective manzacidins A and C syntheses (2002)

See (Scheme 109).

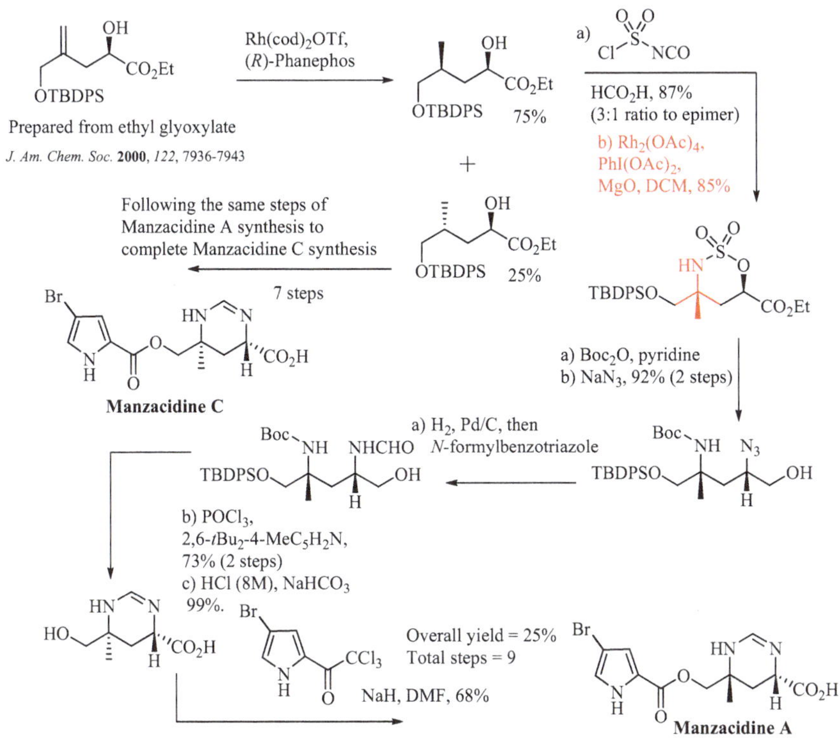


SCHEME 109 Enantioselective manzacidins A and C syntheses.[7]

## 8.3 Mackay's (±)-manzacidin D synthesis (2004)

See (Scheme 110).

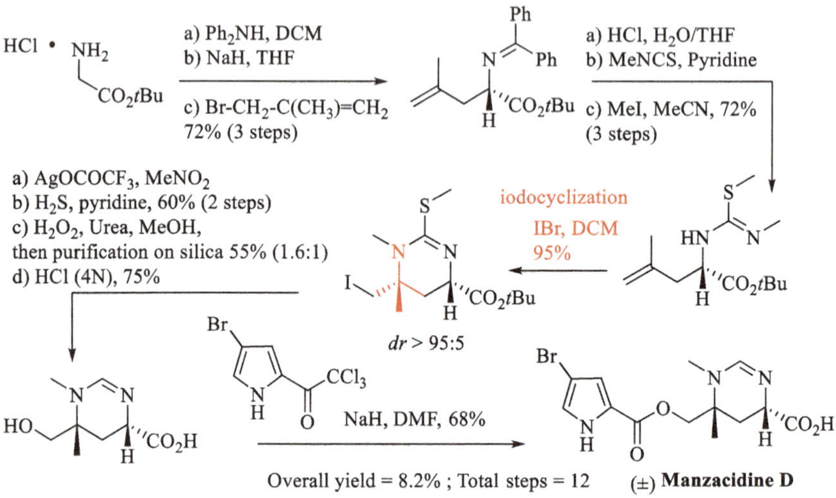


SCHEME 110 Racemic (±)-manzacidin D synthesis.[8]

## 8.4 Lanter's manzacidin C synthesis (2005)

See (Scheme 111).

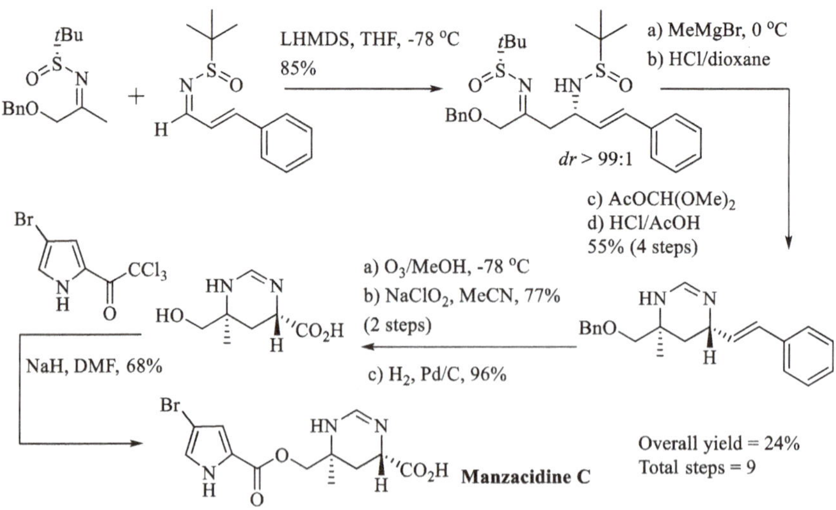


SCHEME 111 Manzacidin C synthesis via aza-Mannich reaction of sulfinimines.[7]

## 8.5 Maruoka's manzacidins A synthesis (2005)

See (Scheme 112).

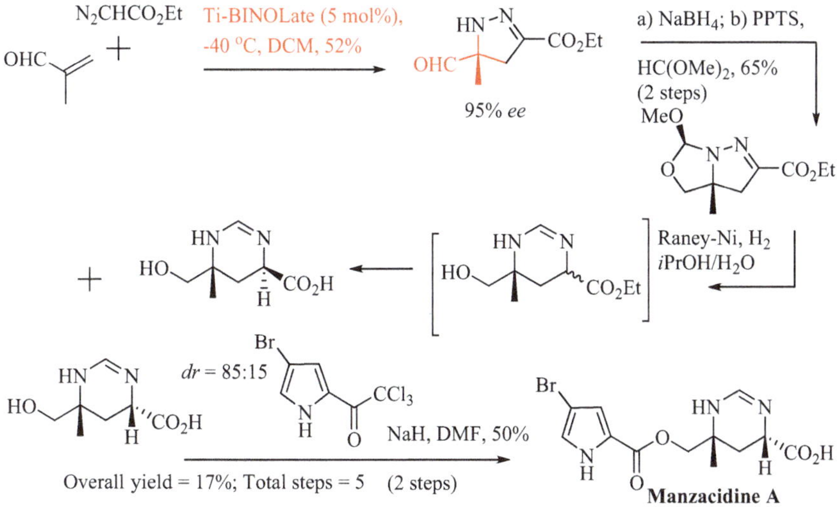


**SCHEME 112** Manzacidins A synthesis via 1,3-dipolar cycloaddition.[9]

## 8.6 Deng's formal synthesis of manzacidin A via tandem conjugate addition–protonation (2006)

See (Scheme 113).

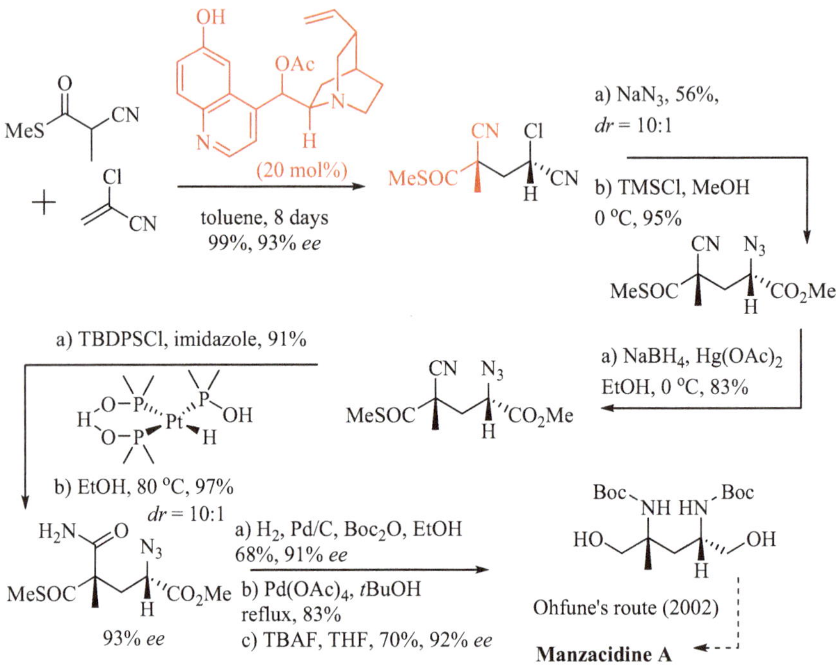


**SCHEME 113** Formal synthesis of manzacidin A via tandem conjugate addition–protonation.[10]

## 8.7 Sibi's manzacidin A synthesis (2007)

See (Scheme 114).

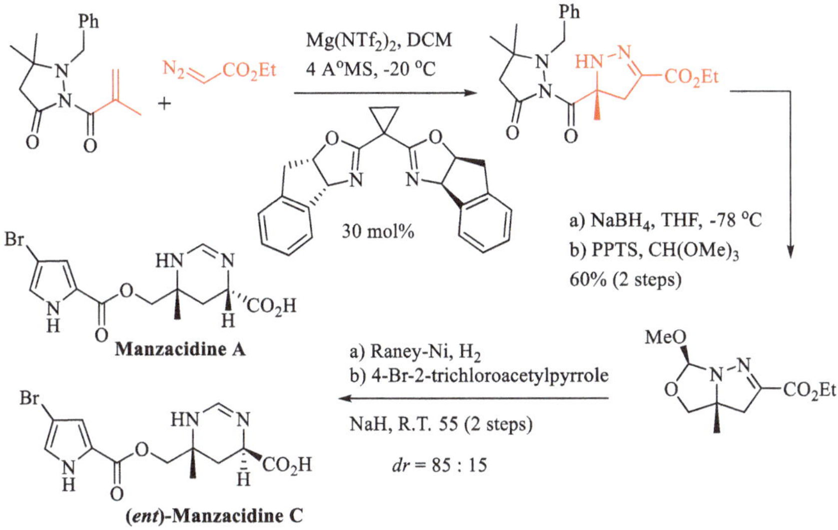


SCHEME 114 Manzacidin A synthesis via 1,3-dipolar cycloaddition.[11]

## 8.8 Ohfune's synthesis of manzacidin B (2007)

See (Scheme 115).

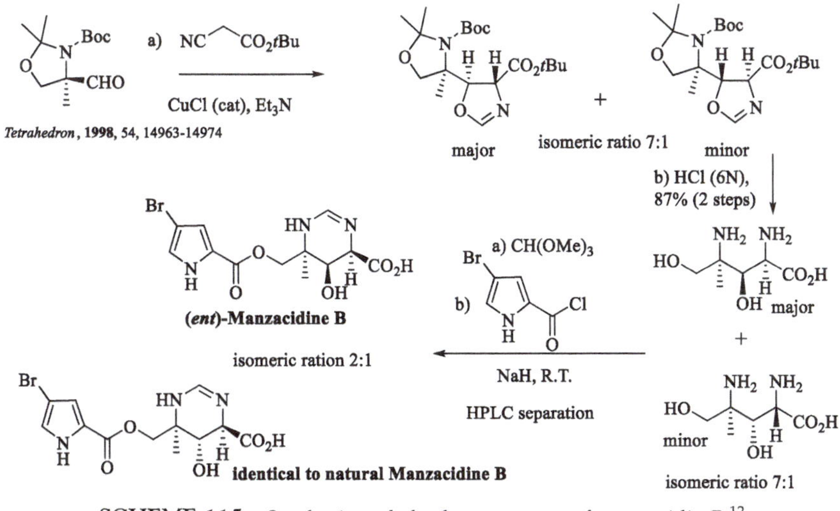


SCHEME 115 Synthesis and absolute structure of manzacidin B.[12]

Later in 2010, Ohfune et al.[13] published corrigendum describing that diasteroisomers (manzacidin B and its *ent* form) are incorrectly assigned. The reinvestigation of study leads to revision of stereochemistry, which demonstrates the enantiomer (*ent* form) as correct structure of manzacidin B (4*R*,5*R*,6*R*) (Scheme 116).

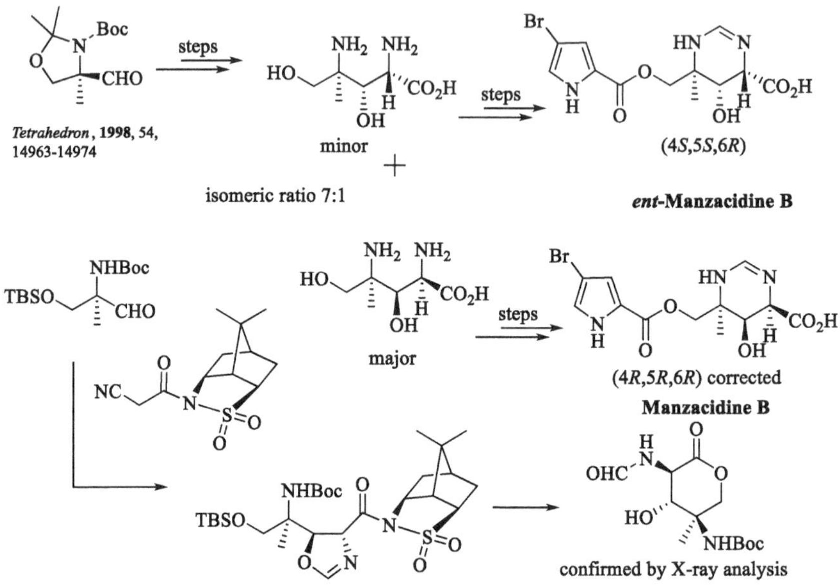


SCHEME 116 Revised synthetic scheme.

## 8.9 Leighton's manzacidin C synthesis (2008)

See (Scheme 117).

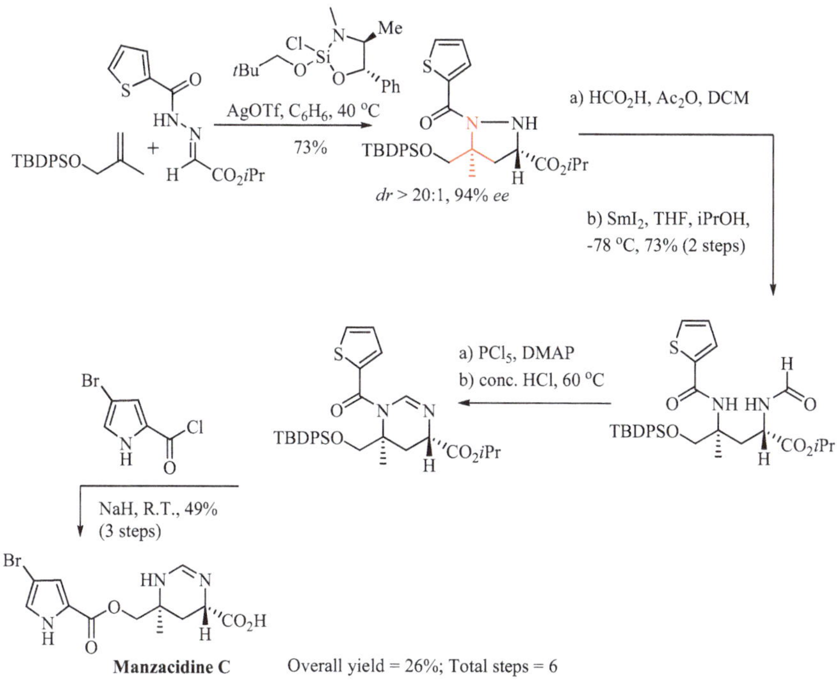


**SCHEME 117** Manzacidin C synthesis via [3 + 2] cycloaddition.[14]

## 8.10 Ohfune's manzacidins A and C synthesis (2008)

See (Scheme 118).

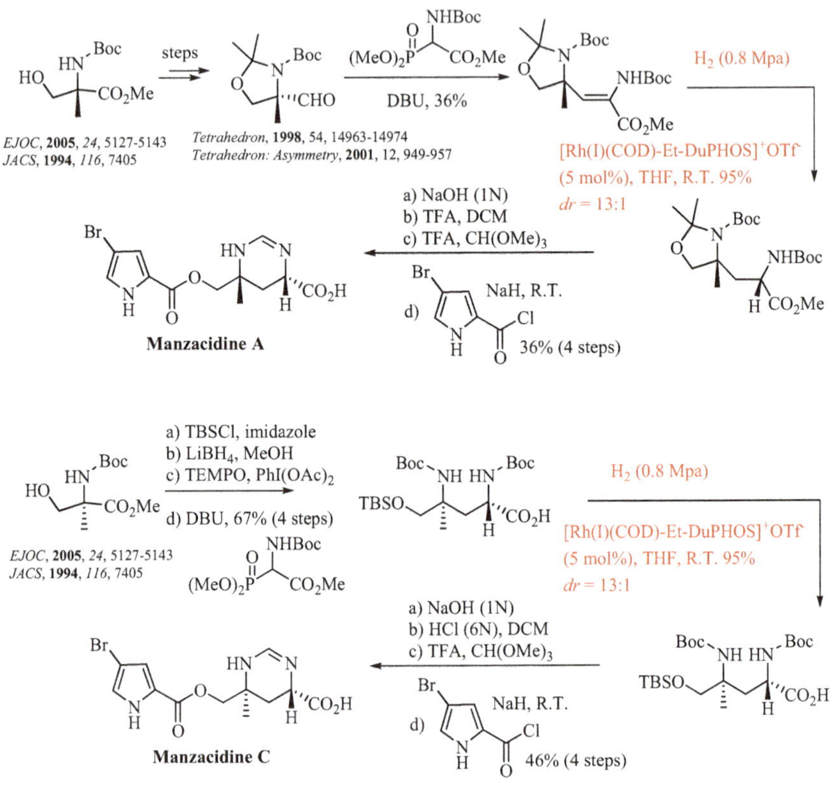


SCHEME 118 Manzacidins A and C synthesis.[12]

# 8.11 Mohapatra's synthesis of manzacidin B (2012)

See (Scheme 119).

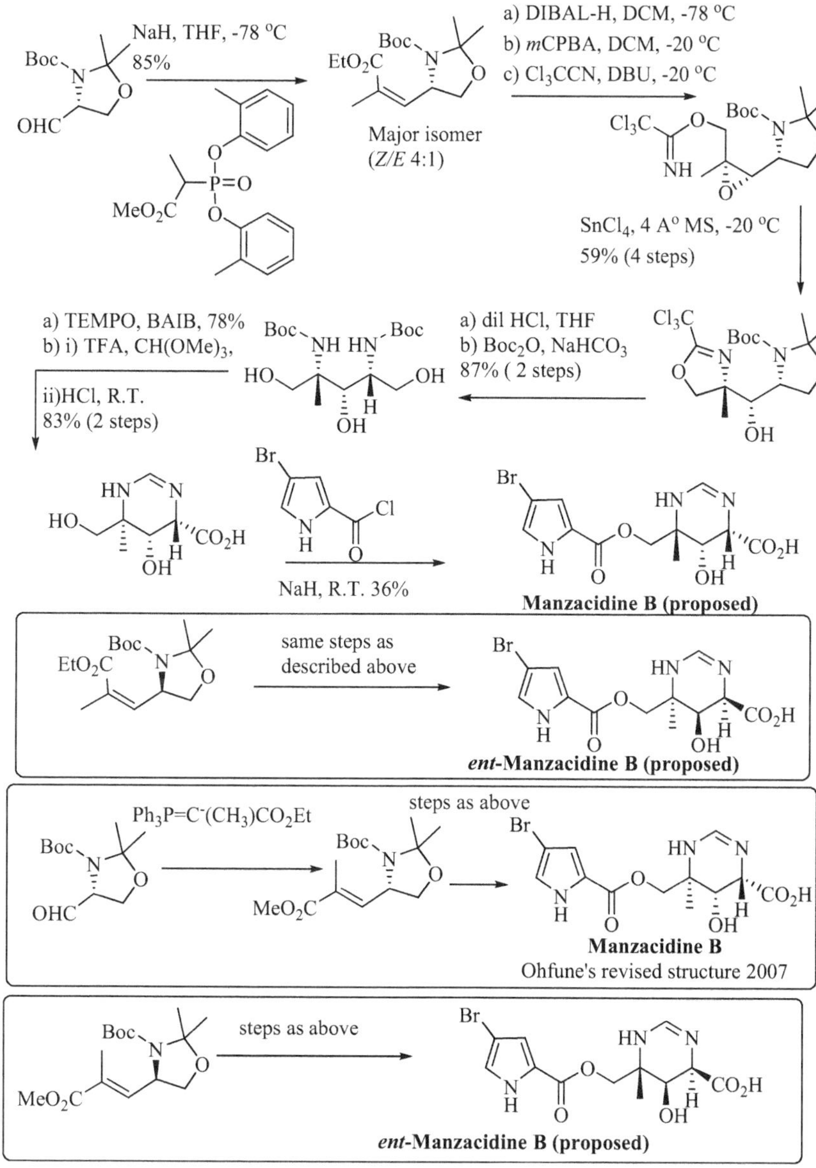


**SCHEME 119** Revised synthesis of manzacidin B.[15]

## 8.12 Ohfune's synthesis of manzacidin B (2012)

See (Scheme 120).

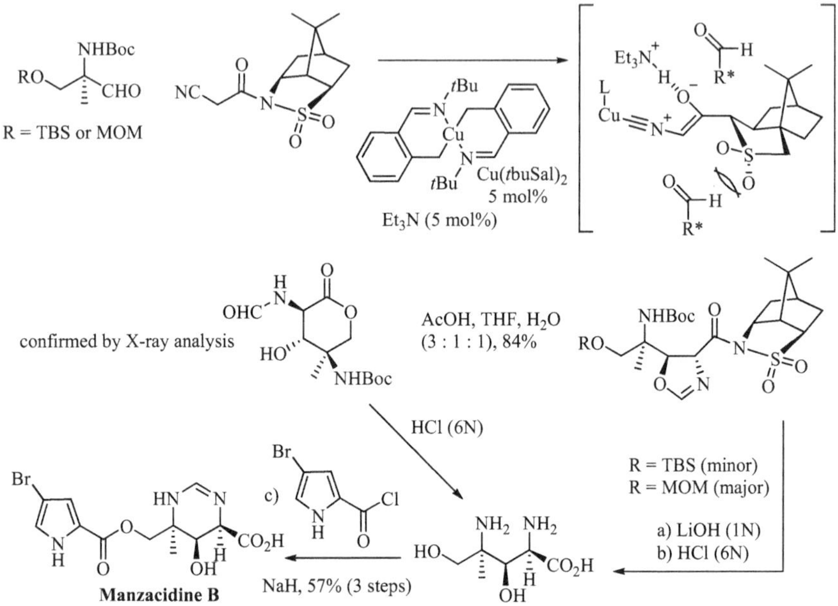


SCHEME 120 Efficient synthesis of manzacidin B.[16]

## 8.13 Kawabata's manzacidin A synthesis (2013)

See (Scheme 121).

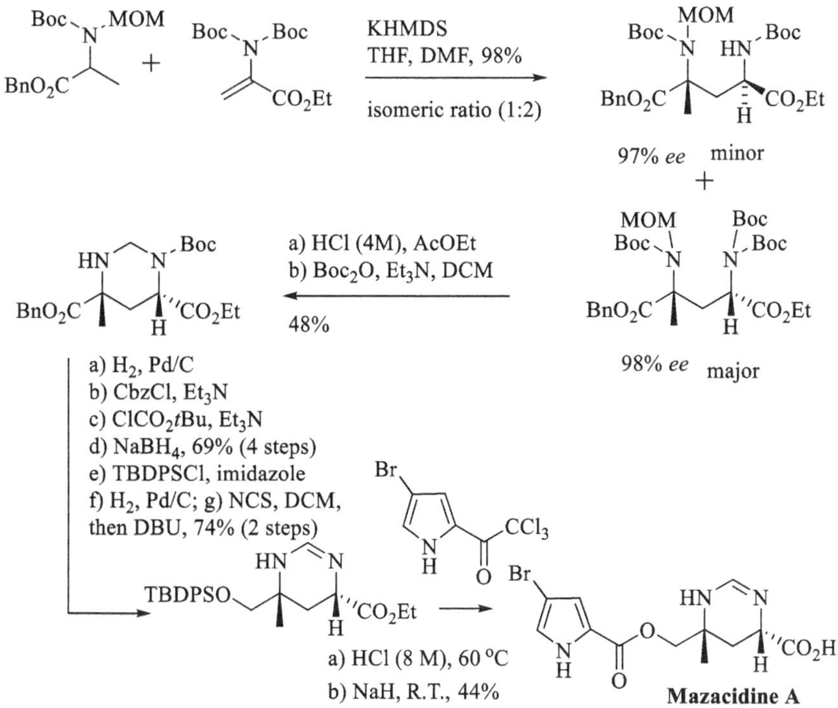


SCHEME 121 Manzacidin A synthesis asymmetric intermolecular conjugate.[17]

## 8.14 Ichikawa's manzacidins A and C synthesis (2012)

See (Scheme 122).

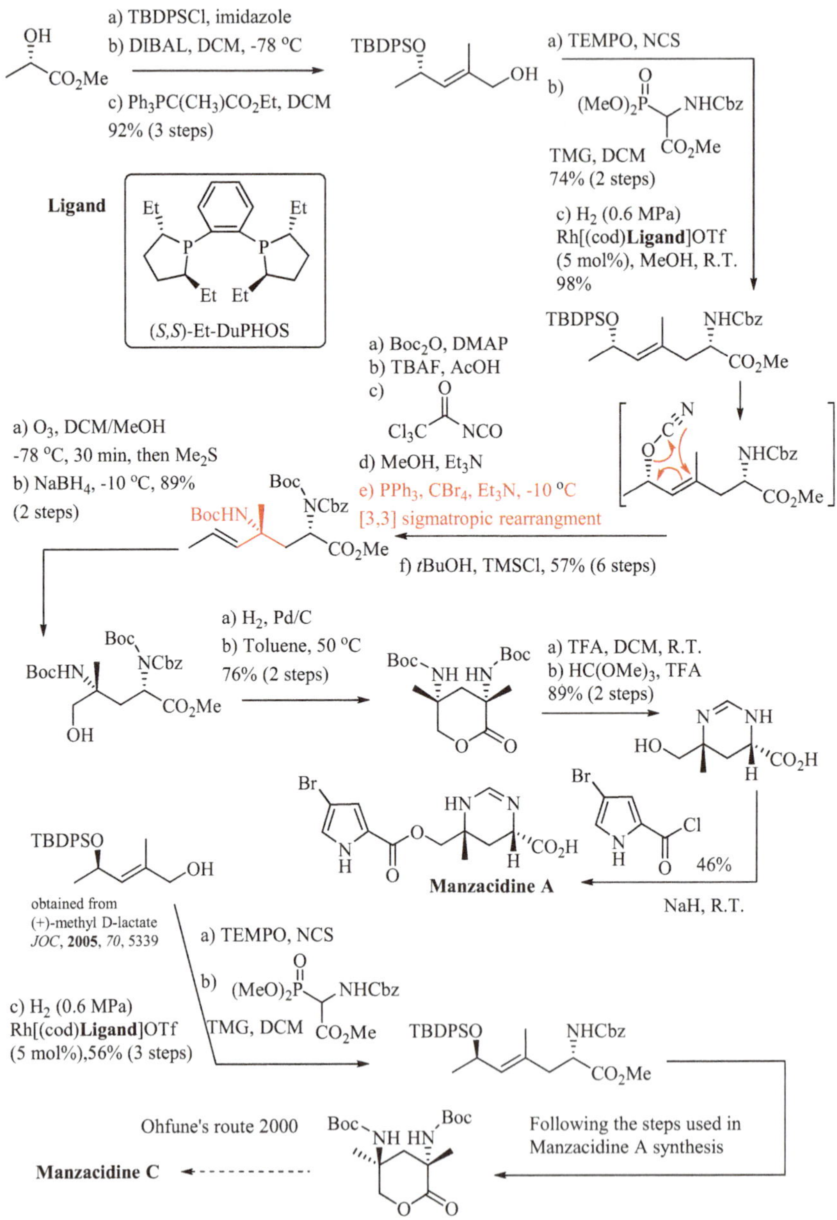


SCHEME 122 Efficient synthesis of manzacidins A and C.[18]

## 8.15 Inoue's manzacidin A synthesis (2015)

See (Scheme 123).

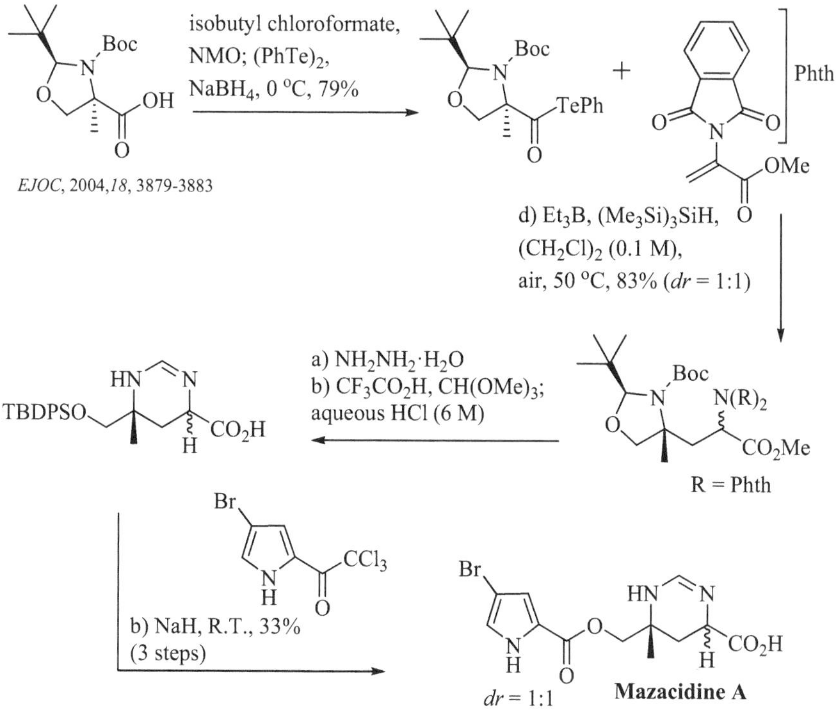


**SCHEME 123** Manzacidin A synthetic pathway.[19]

## 8.16 Sakakura's synthesis of mazacidins A and C (2017)

See (Scheme 124).

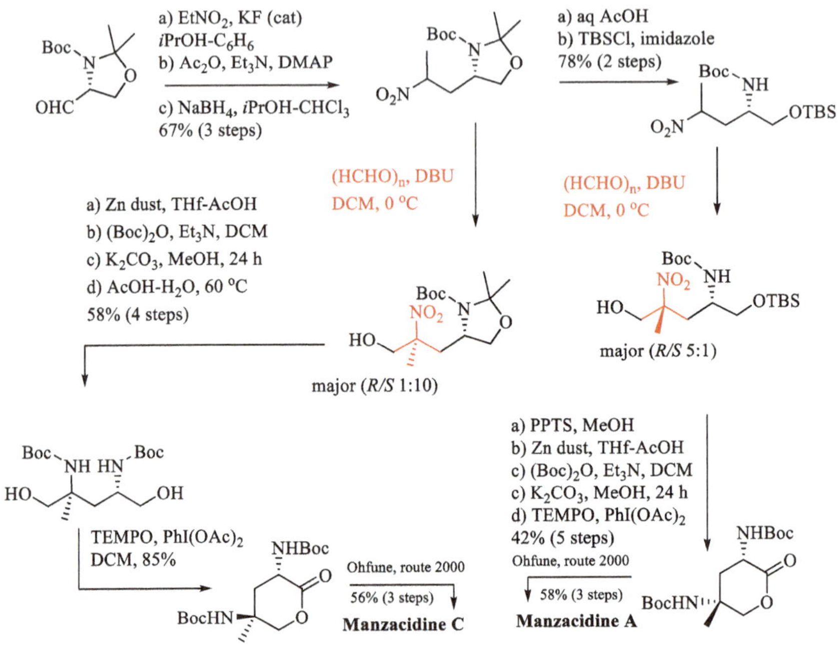


SCHEME 124 Mazacidins A and C via Henry reaction.[20]

## 8.17 Ukaji's formal synthesis of manzacidin (2017)

See (Scheme 125).

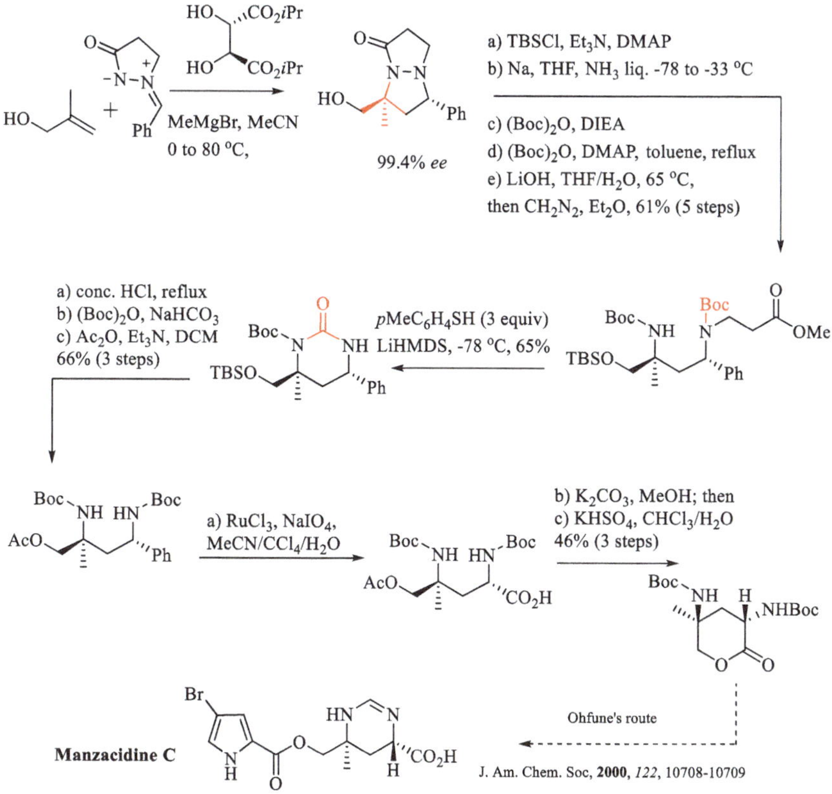


SCHEME 125 Manzacidin C via 1,3-dipolar cycloaddition of azomethine imines.[21]

## 8.18 Renata's formal synthesis of manzacidin C (2018)

See (Scheme 126).

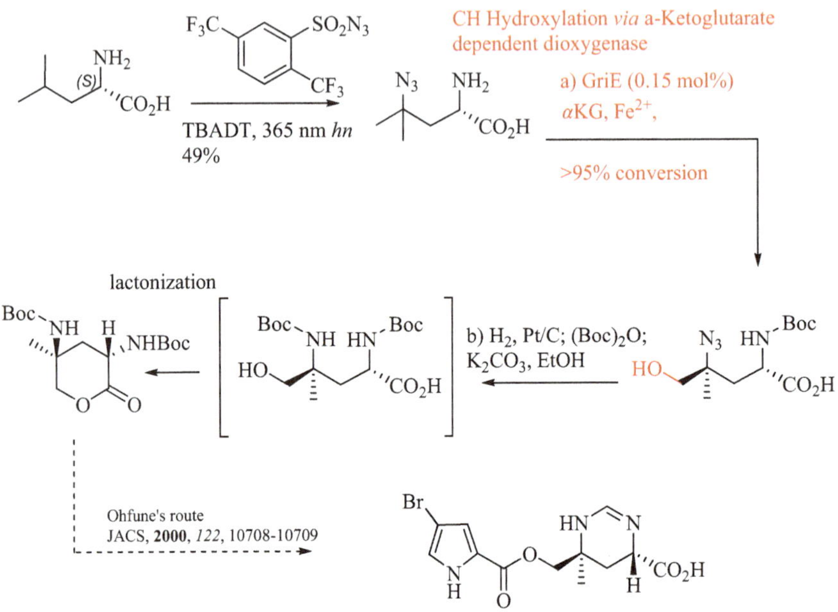


SCHEME 126 Manzacidin C via GriE catalyzed hydroxylation.[22]

# References


1. Kobayashi, J.; Kanda, F.; Ishibashi, M.; Shigemori, H.; Manzacidins, A. C. Novel Tetrahydropyrimidine Alkaloids from the Okinawan Marine Sponge Hymeniacidon Sp. *J. Org. Chem.* **1991,** *56,* 4574–4576.
2. Jahn, T.; König, G. M.; Wright, A. D.; Wörheide, G.; Reitner, J.; Manzacidin, D. An Unprecedented Secondary Metabolite from the "Living Fossil" Sponge Astrosclera Willeyana. *Tetrahedron Lett.* **1997,** *38,* 3883–3884.
3. Faulkner, D. J. Marine Natural Products. *Nat. Prod. Rep.* **1998,** *15,* 113–158.
4. König, G. M.; Wright, A. D.; Franzblau, S. G. Assessment of Antimycobacterial Activity of a Series of Mainly Marine Derived Natural Products. *Planta Med.* **2000,** *66,* 337–342.
5. Hashimoto, T.; Maruoka, K. Syntheses of Manzacidins: a Stage for the Demonstration of Synthetic Methodologies. *Org. Biomol. Chem.* **2008,** *6,* 829–835.
6. Namba, K.; Shinada, T.; Teramoto, T.; Ohfune, Y. Total Synthesis and Absolute Structure of Manzacidin A and C. *J. Am. Chem. Soc.* **2000,** *122,* 10708–10709.
7. Wehn, P. M.; Du Bois, J. Enantioselective Synthesis of the Bromopyrrole Alkaloids Manzacidin A and C by Stereospecific C–H Bond Oxidation. *J. Am. Chem. Soc.* **2002,** *124* (44), 12950–12951.
8. Drouin, C.; Woo, J. C.; MacKay, D. B.; Lavigne, R. M. Total Synthesis of (±)-manzacidin D. *Tetrahedron Lett.* **2004,** *45* (39), 7197–7199.

9. Kano, T.; Hashimoto, T.; Maruoka, K. Enantioselective 1, 3-dipolar Cycloaddition Reaction between Diazoacetates and α-substituted Acroleins: Total Synthesis of Manzacidin A. *J. Am. Chem. Soc.* **2006,** *128,* 2174–2175.
10. Wang, Y.; Liu, X.; Deng, L. Dual-function Cinchona Alkaloid Catalysis: Catalytic Asymmetric Tandem Conjugate Addition– Protonation for the Direct Creation of Nonadjacent Stereocenters. *J. Am. Chem. Soc.* **2006,** *128,* 3928–3930.
11. Sibi, M. P.; Stanley, L. M.; Soeta, T. Enantioselective 1, 3-dipolar Cycloadditions of Diazoacetates with Electron-Deficient Olefins. *Org. Lett.* **2007,** *9,* 1553–1556.
12. Shinada, T.; Ikebe, E.; Oe, K.; Namba, K.; Kawasaki, M.; Ohfune, Y. Synthesis and Absolute Structure of Manzacidin B. *Org. Lett.* **2007,** *9,* 1765–1767.
13. Shinada, T.; Ikebe, E.; Oe, K.; Namba, K.; Kawasaki, M.; Ohfune, Y. Synthesis and Absolute Structure of Manzacidin B. *Org. Lett.* **2010,** *12,* 2170-1767.
14. Tran, K.; Lombardi, P. J.; Leighton, J. L. An Efficient Asymmetric Synthesis of Manzacidin C. *Org. Lett.* **2008,** *10,* 3165–3167.
15. Sankar, K.; Rahman, H.; Das, P. P.; Bhimireddy, E.; Sridhar, B.; Mohapatra, D. K. Practical Syntheses of Proposed and Revised Manzacidin B and Their Congeners. *Org. Lett.* **2012,** *14,* 1082–1085.
16. Shinada, T.; Oe, K.; Ohfune, Y. Efficient Total Synthesis of Manzacidin B. *Tetrahedron Lett.* **2012,** *53,* 3250–3253.
17. Yoshimura, T.; Kinoshita, T.; Yoshioka, H.; Kawabata, T. Asymmetric Intermolecular Conjugate Addition of Amino Acid Derivatives via Memory of Chirality: Total Synthesis of Manzacidin A. *Org. Lett.* **2013,** *15,* 864–867.
18. Ichikawa, Y.; Okumura, K.; Matsuda, Y.; Hasegawa, T.; Nakamura, M.; Fujimoto, A.; Masuda, T.; Nakano, K.; Kotsuki, H. Synthesis of Manzacidin A and C: Efficient Construction of Quaternary Carbon Stereocenters Bearing Nitrogen Substituents. *Org. Biomol. Chem.* **2012,** *10,* 614–622.
19. Nagatomo, M.; Nishiyama, H.; Fujino, H.; Inoue, M. Decarbonylative Radical Coupling of α-Aminoacyl Tellurides: Single-Step Preparation of γ-Amino and α, β-Diamino Acids and Rapid Synthesis of Gabapentin and Manzacidin A. *Angew. Chem.* **2015,** *127,* 1557–1561.
20. Kudoh, T.; Araki, Y.; Miyoshi, N.; Tanioka, M.; Sakakura, A. Diastereodivergent Henry Reaction for the Stereoselective Construction of Nitrogen-Containing Tetrasubstituted Carbons: Application to Total Synthesis of Manzacidins A and C. *Asian J. Org. Chem.* **2017,** *6,* 1760–1763.
21. Tong, T. M. T.; Soeta, T.; Suga, T.; Kawamoto, K.; Hayashi, Y.; Ukaji, Y. Formal Total Synthesis of Manzacidin C Based on Asymmetric 1, 3-Dipolar Cycloaddition of Azomethine Imines. *J. Org. Chem.* **2017,** *82,* 1969–1976.
22. Zwick, C. R., III; Renata, H.; Remote, C.–H. Hydroxylation by an α-Ketoglutarate-Dependent Dioxygenase Enables Efficient Chemoenzymatic Synthesis of Manzacidin C and Proline Analogs. *J. Am. Chem. Soc.* **2018,** *140* (3), 1165–1169.

# Neooxazolomycin

Neooxazolomycin and oxazolomycin A are most well-known members of oxazolomycin natural product family. The compounds were isolated by Uemura's group in 1985 from the strains of *Streptomyces*.[1] The oxazolomycin family showed in vitro antiviral activity,[2] in vivo anticancer activity against p-388 leukemia,[3] and broad range antibacterial activity.[4] All the members of oxazolomycin family have closely related structure bearing (1) pyroglutamate core (right), (2) dialkene chain (middle) and (3) oxazole-linked fragment (left). The neooxazolomycin has $\gamma$-lactone in fused bicyclic system in contrast to oxazolomycin, which possessed spirocyclic consisting of $\beta$-lactone and $\gamma$-lactone rings. The key $\alpha,\alpha$-dialkyl $\alpha$-amino acid framework in both types of molecules has been constructed via different methodologies in the their synthetic pathways (Fig. 9).

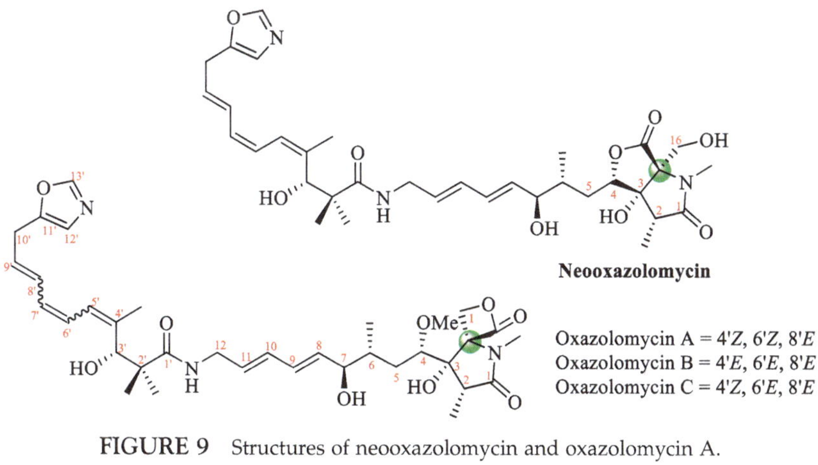


FIGURE 9 Structures of neooxazolomycin and oxazolomycin A.


https://doi.org/10.1016/B978-0-12-822262-1.00009-3

## 9.1 Kende's first enantioselective total neooxazolomycin synthesis (1990)

See (Scheme 127).

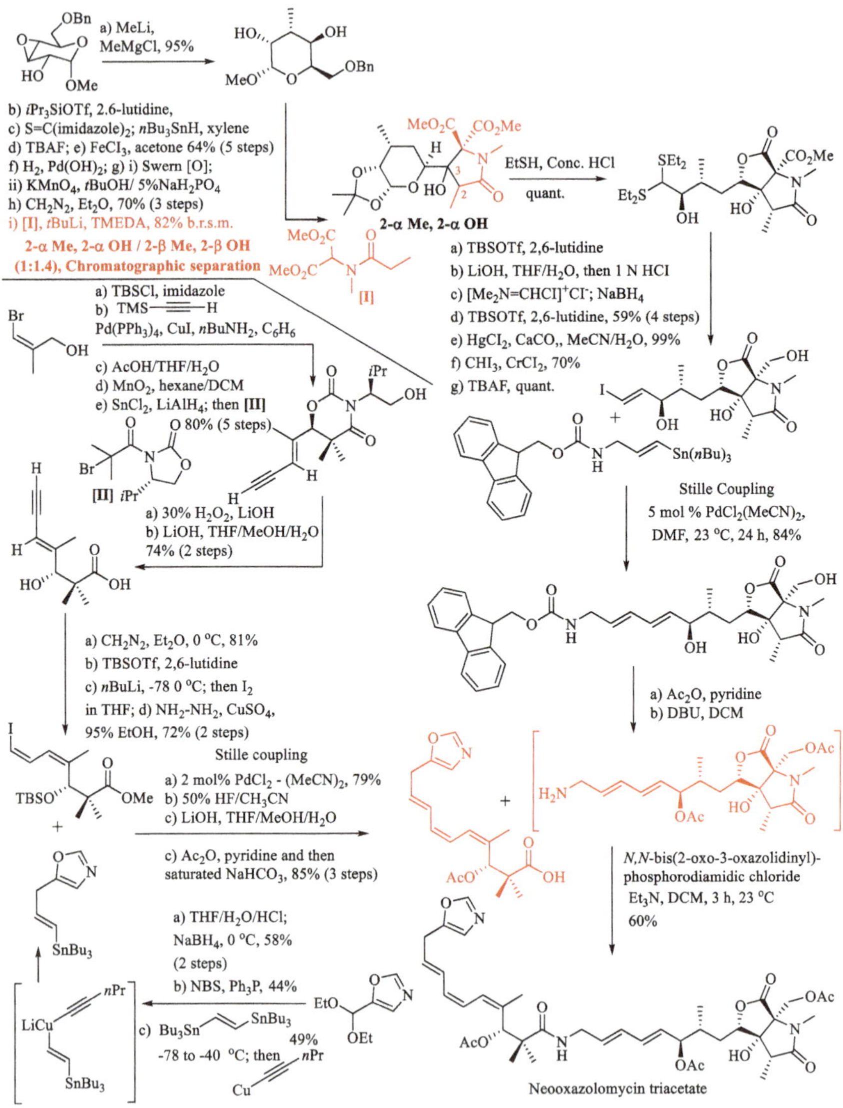


**SCHEME 127** First enantioselective total neooxazolomycin synthesis.[5]

## 9.2 Hatakeyama's total neooxazolomycin synthesis (2007)

See (Scheme 128).

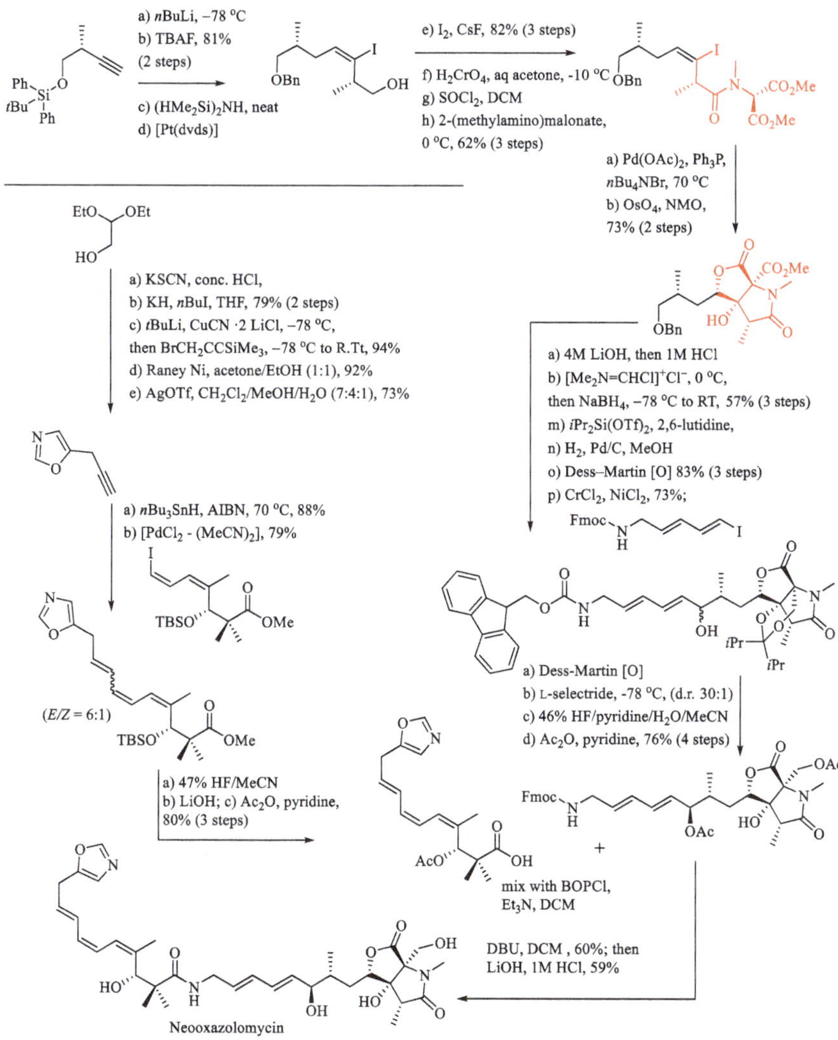


SCHEME 128 Neooxazolomycin total synthesis.[6]

## 9.3 Hatakeyama's total oxazolomycin synthesis (2011)

See (Scheme 129).

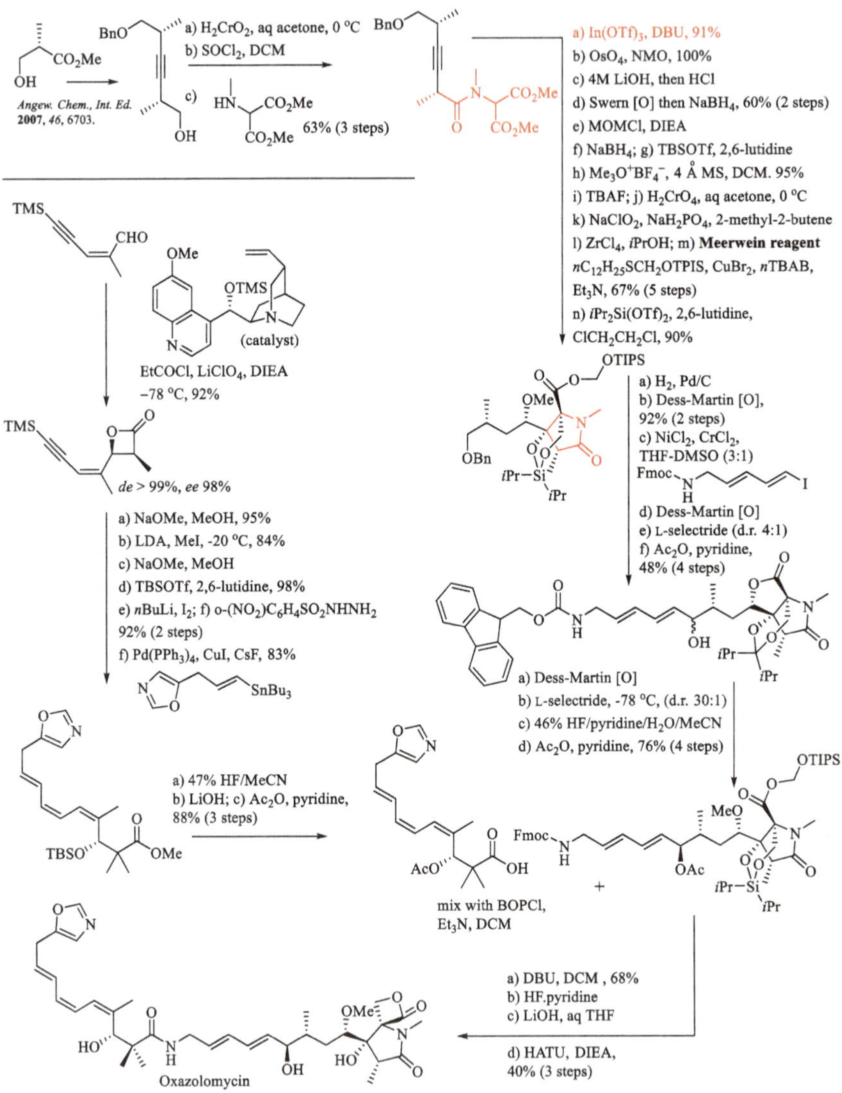


Total synthesis with 1.4% overall yield in 34 steps from methyl (*S*)-3-hydroxy-2-methylpropionate.

SCHEME 129 Total oxazolomycin synthesis.[7]

## 9.4 Pattenden's approach toward oxazolomycin A and neooxazolomycin synthesis (2007)

The advanced intermediate could be used as precursor to construct the lactone–pyrrolidinone ring systems in oxazolomycin A and neo-oxazolomycin synthesis (Scheme 130).

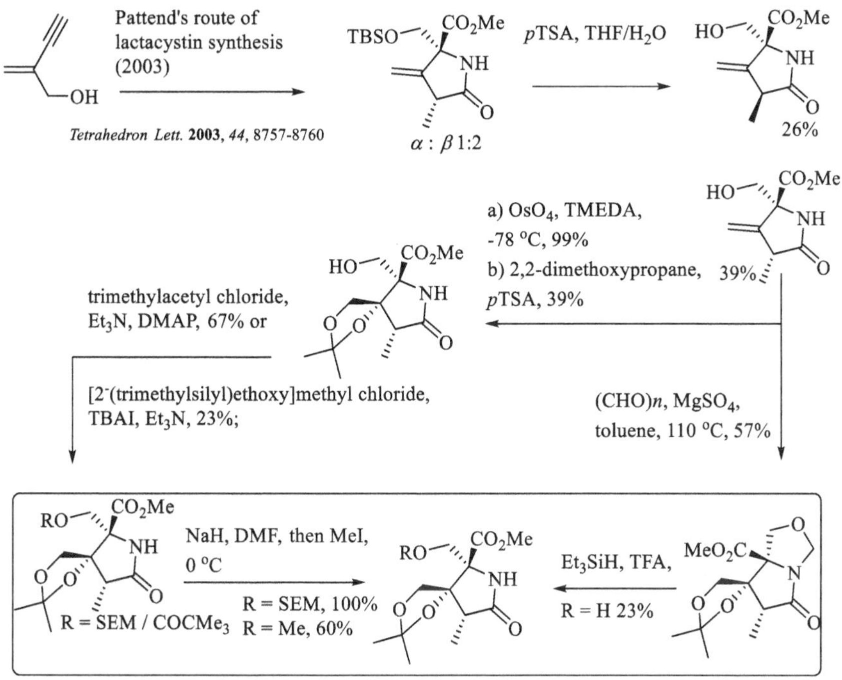


SCHEME 130 Synthetic study toward oxazolomycin A and neooxazolomycin synthesis.[8]

## 9.5 Moloney's approach toward oxazolomycin (2002)

See (Scheme 131).

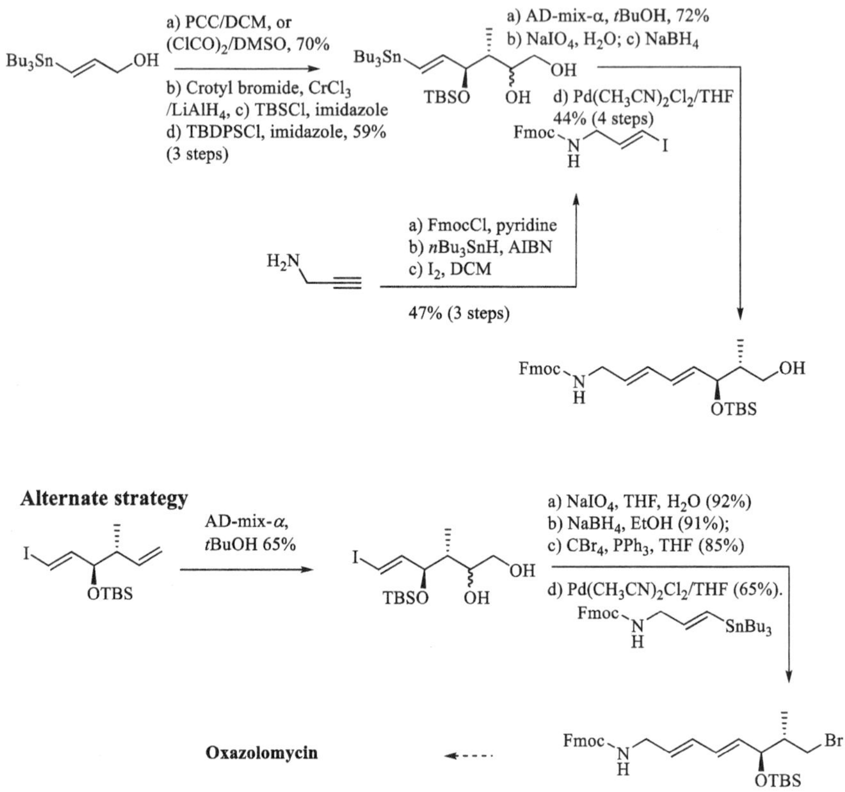


SCHEME 131 Synthesis of middle fragment of oxazolomycin.[9]

# 9.6 Taylor's formal synthesis of (+)-neooxazolomycin (2011)

See (Scheme 132).

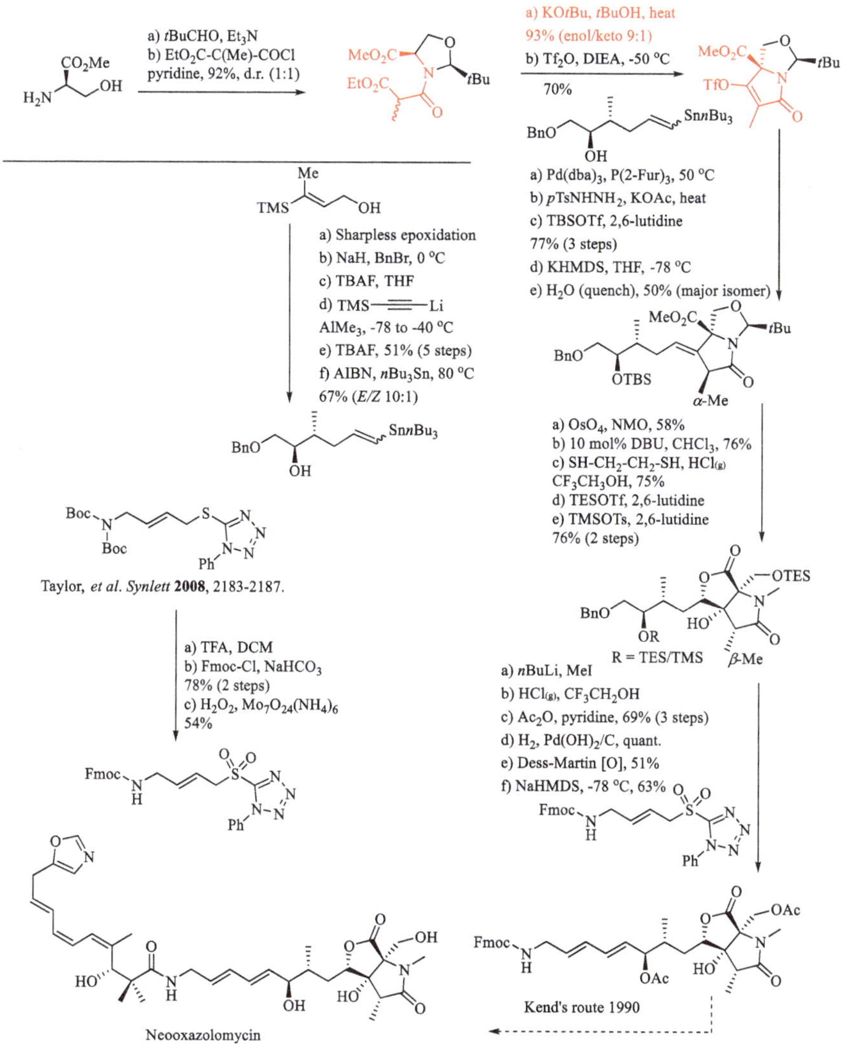


SCHEME 132 Formal synthesis of (+)-neooxazolomycin.[10]

## 9.7 Mohapatra's approach toward oxazolomycin (2006)

See (Scheme 133).

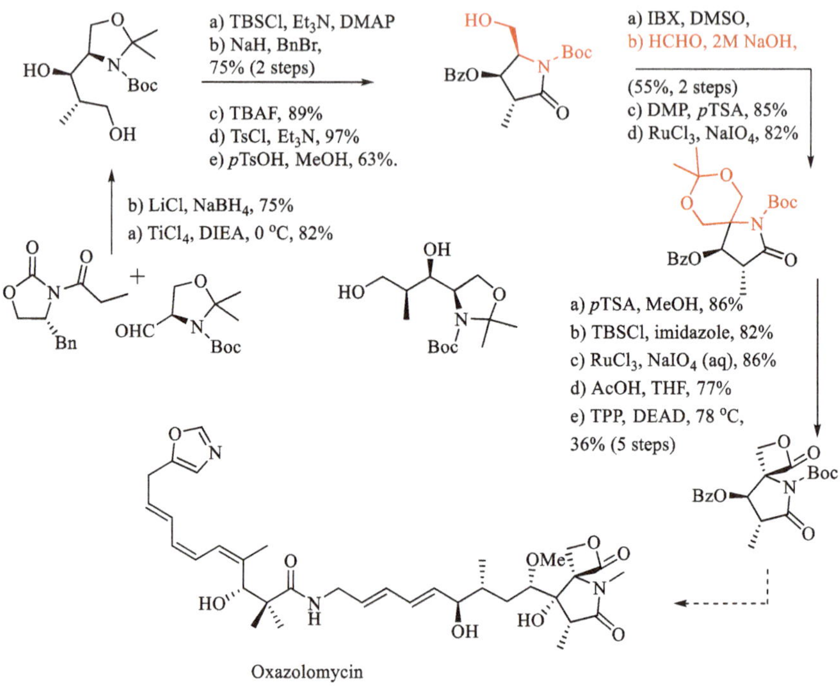


SCHEME 133 Approach toward spirocyclic system of oxazolomycin.[10]

# 9.8 Donohoe's approach toward pyrrolidinone core of oxazolomycin A (2012)

See (Scheme 134).

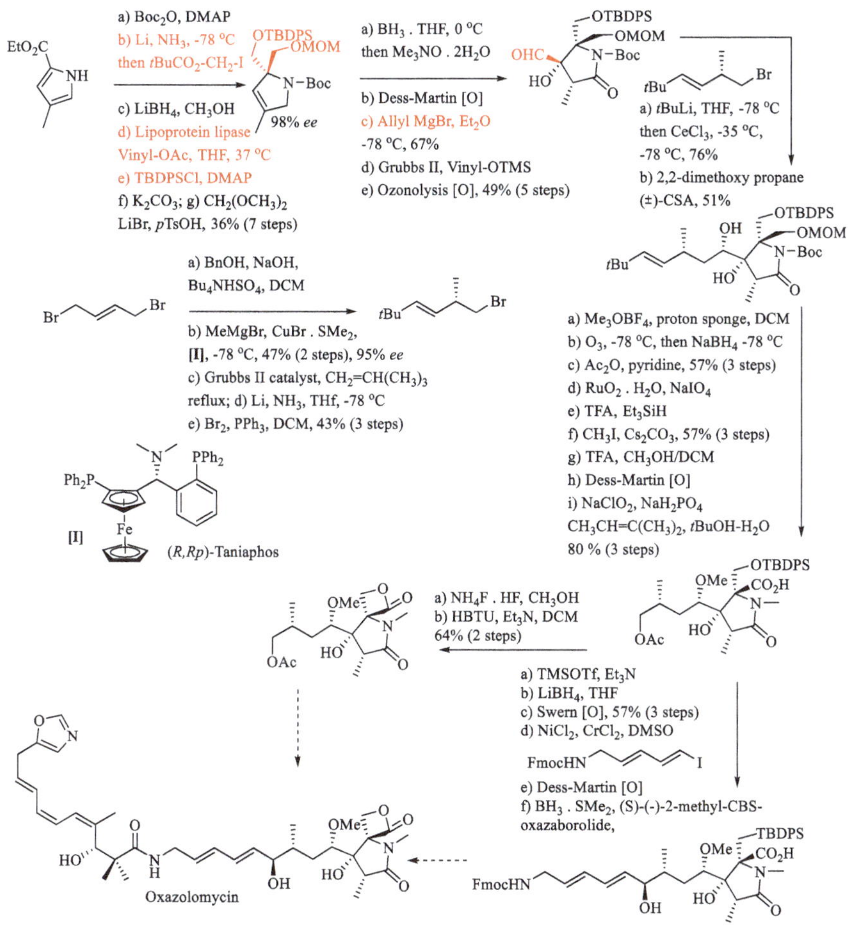


SCHEME 134 Synthetic studies Toward pyrrolidinone core of oxazolomycin A.[11]

# References

1. Mori, T.; Takahashi, K.; Kashiwabara, M.; Uemura, D.; Katayama, C.; Iwadare, S.; Shizuri, Y.; Mitomo, R.; Nakano, F.; Matsuzaki, A. Structure of Oxazolomycin, a Novel $\beta$-lactone Antibiotic. *Tetrahedron Lett.* **1985,** *26,* 1073–1076.
2. Tonew, E.; Tonew, M.; Gräfe, U.; Zöpel, P. On the Antiviral Activity of Diffusomycin (Oxazolomycin). *Acta Virol.* **1992,** *36,* 166–172.
3. Takahashi, K.; Kawabata, M.; Uemura, D.; Iwadare, S.; Mitomo, R.; Nakano, F.; Matsuzaki, A. Structure of Neooxazolomycin, an Antitumor Antibiotic. *Tetrahedron Lett.* **1985,** *26,* 1077–1078.
4. Zhao, C.; Coughlin, J. M.; Ju, J.; Zhu, D.; Wendt-Pienkowski, E.; Zhou, X.; Wang, Z.; Shen, B.; Deng, Z. Oxazolomycin Biosynthesis in Streptomyces Albus JA3453 Featuring an "Acyltransferase-less" Type I Polyketide Synthase that Incorporates Two Distinct Extender Units. *J. Biol. Chem.* **2010,** *285,* 20097–20108.
5. Kende, A. S.; Kawamura, K.; DeVita, R. J. Enantioselective Total Synthesis of Neooxazolomycin. *J. Am. Chem. Soc.* **1990,** *112,* 4070–4072.
6. Onyango, E. O.; Tsurumoto, J.; Imai, N.; Takahashi, K.; Ishihara, J.; Hatakeyama, S. Total Synthesis of Neooxazolomycin. *Angew. Chem. Int. Ed.* **2007,** *46,* 6703–6705.
7. Eto, K.; Yoshino, M.; Takahashi, K.; Ishihara, J.; Hatakeyama, S. Total Synthesis of Oxazolomycin A. *Org. Lett.* **2011,** *13,* 5398–5401.
8. Bennett, N. J.; Prodger, J. C.; Pattenden, G. A Synthesis of a Common Intermediate to the Lactone–Pyrrolidinone Ring Systems in Oxazolomycin A and Neooxazolomycin. *Tetrahedron* **2007,** *63,* 6216–6231.
9. Wang, Z.; Moloney, M. G. Synthesis of the Middle Fragment of Oxazolomycin. *Tetrahedron Lett.* **2002,** *43,* 9629–9632.
10. Bastin, R.; Dale, J. W.; Edwards, M. G.; Papillon, J. P.; Webb, M. R.; Taylor, R. J. Formal Synthesis of (+)-neooxazolomycin *via* a Stille Cross-Coupling/deconjugation Route. *Tetrahedron* **2011,** *67,* 10026–10044.
11. Donohoe, T. J.; O'Riordan, T. J.; Peifer, M.; Jones, C. R.; Miles, T. J. Asymmetric Synthesis of the Fully Elaborated Pyrrolidinone Core of Oxazolomycin A. *Org. Lett.* **2012,** *14,* 5460–5463.

CHAPTER

# 10

# Sphingofungins

## 10.1 Abstract

Sphingofungins are a family of natural products with potent antifungal properties, isolated by Merck group from the fermentation of *Paecilomyces variotii* in 1992.[1] The sphingofungins possess unique polyhydroxyamino acid moiety and long lipid chain and are like sphingosine compounds that inhibit serine palmitoyltransferase, an enzyme used in the initial step of sphingolipid biosynthesis.[2,3] The sphingofungins E and F bear quaternary center along with trans double bond in polyhydroxy amine scaffold and have antifungal activities against several human pathogenic fungi. The significant biological properties with interesting structural features, i.e., presence of α-substituted α-amino acid with trihydroxy alkylated chain make them interesting target for synthesis chemists (Fig. 10).

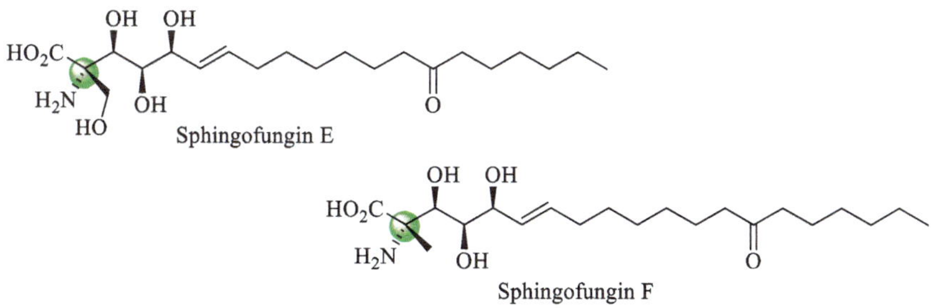


FIGURE 10 Structures of sphingofungins (E and F).


https://doi.org/10.1016/B978-0-12-822262-1.00010-X

## 10.2 Kobayashi's asymmetric synthesis of sphingofungin F (1998)

See (Scheme 135).

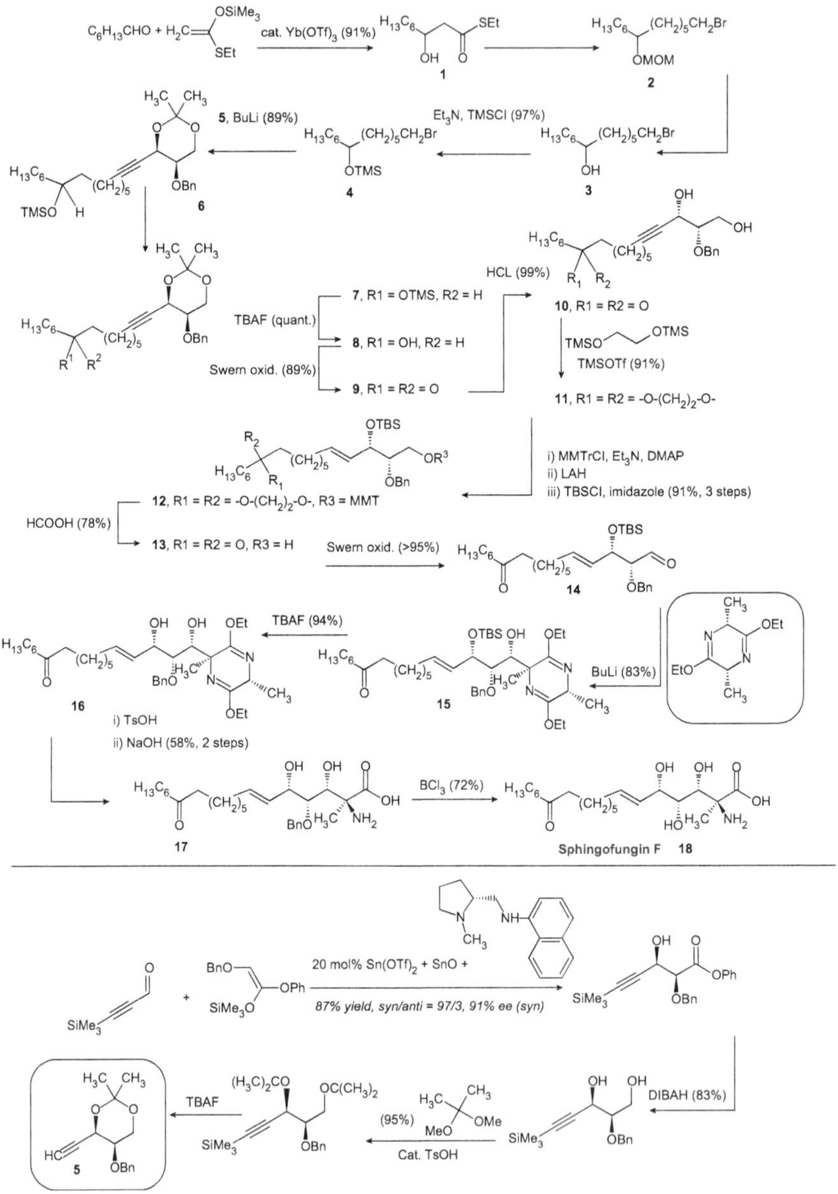


SCHEME 135 Asymmetric synthesis of sphingofungin F.[4]

## 10.3 Trost's total synthesis of sphingofungin F (1998)

See (Scheme 136).

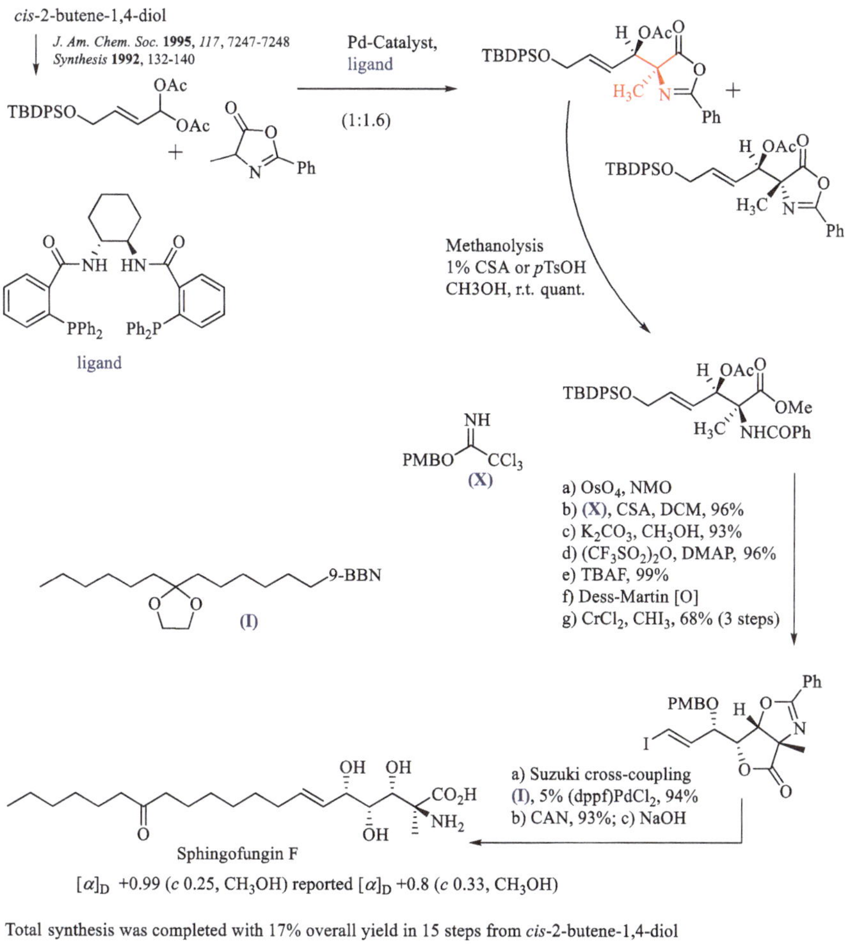


SCHEME 136 Total synthesis of sphingofungin F.[5]

## 10.4 Trost's total synthesis of sphingofungin F (2001)

See (Scheme 137).

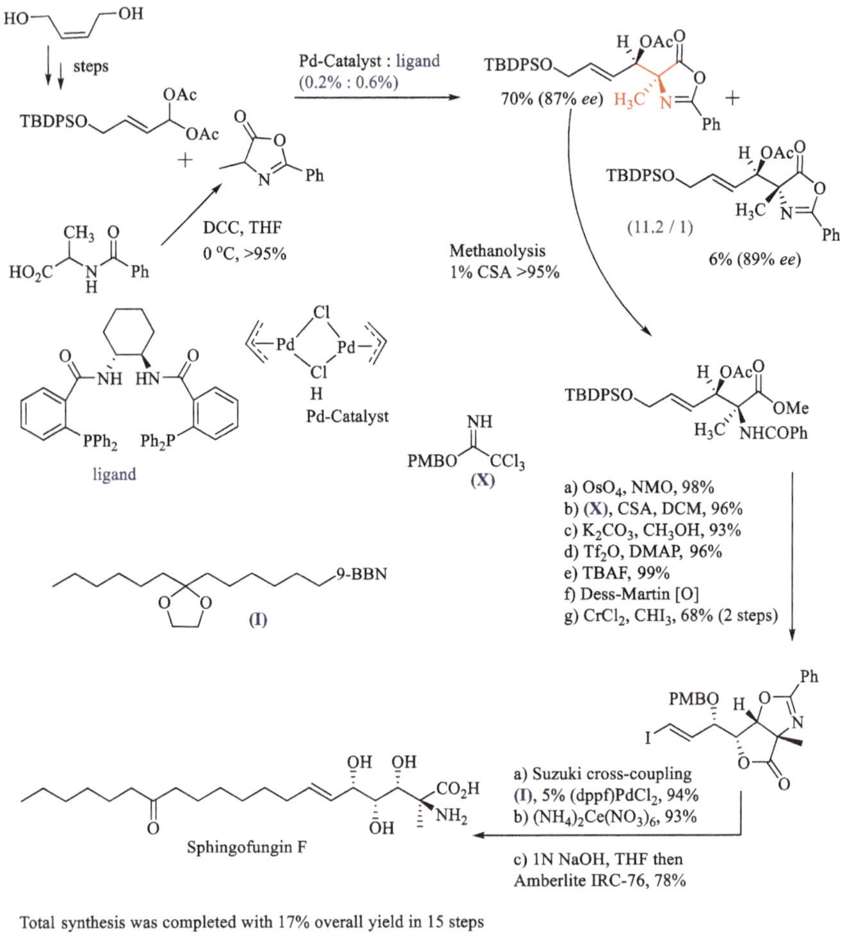


SCHEME 137 Total synthesis of sphingofungin F.[6]

## 10.5 Trost's total synthesis of sphingofungin E (2001)

See (Scheme 138).

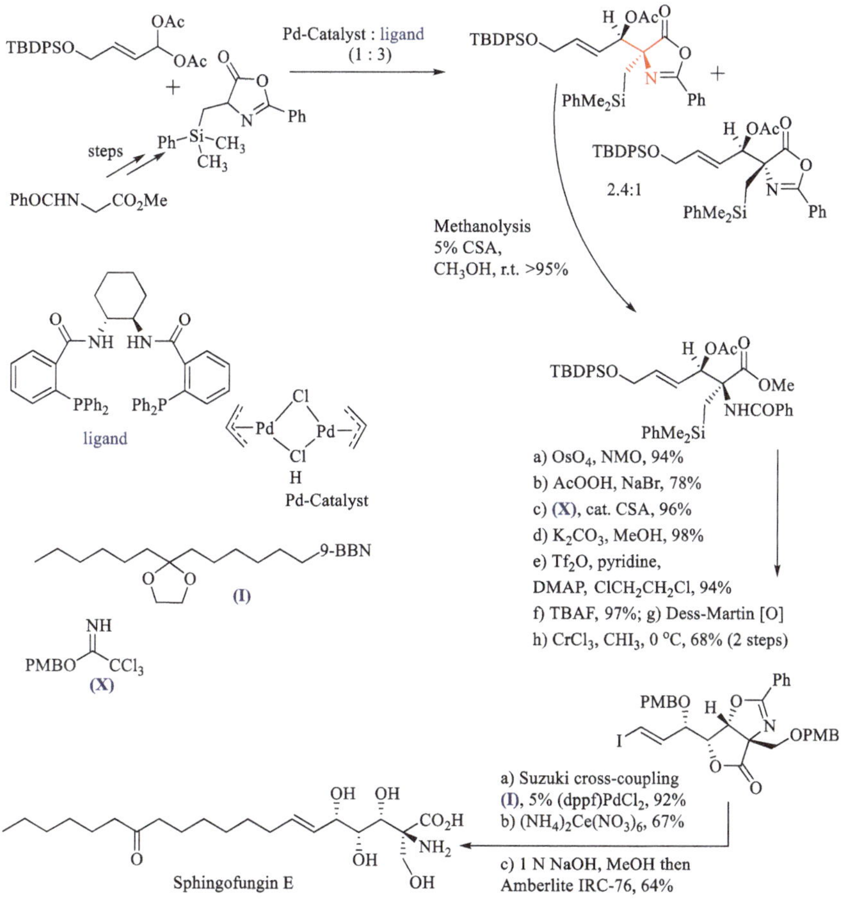


Total synthesis was completed in 17 steps with 5.1% overall yield

**SCHEME 138** Total synthesis of sphingofungin E.[6]

## 10.6 Lin's total synthesis of sphingofungin F (2000)

See (Scheme 139).

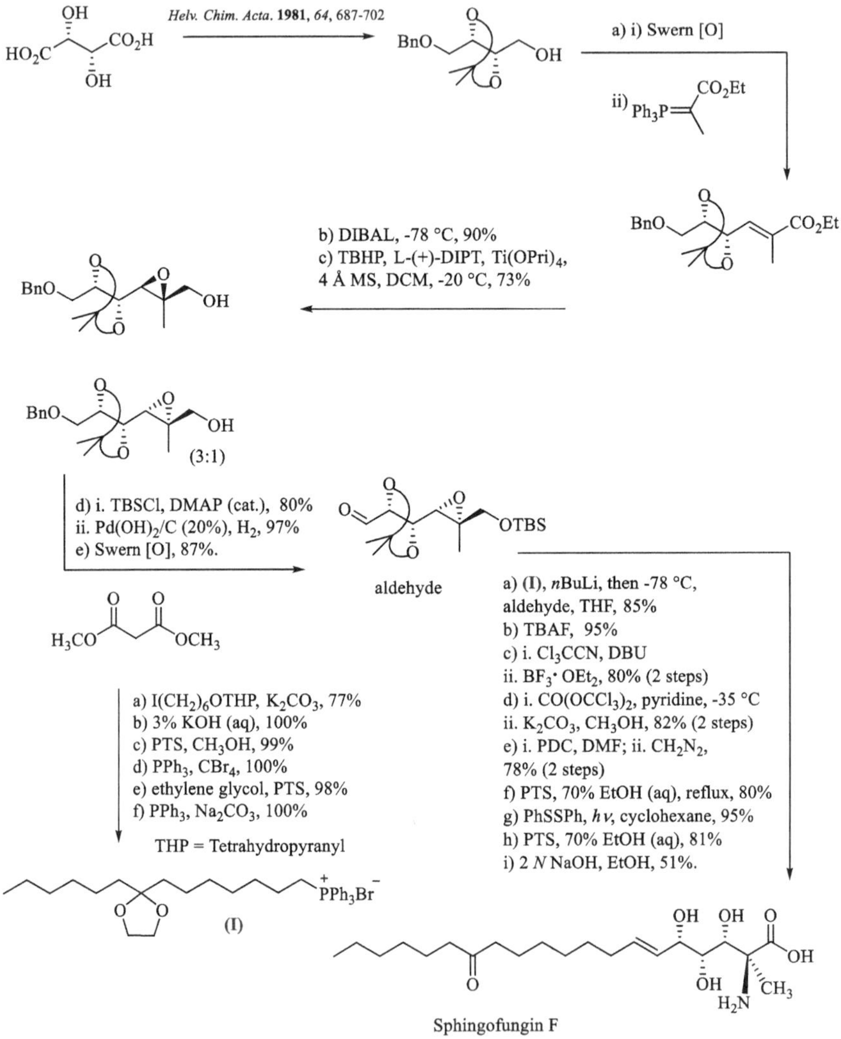


Total synthesis was completed with 3.7% overall yield in 22 steps from L-(+)-tartaric acid

SCHEME 139 Total synthesis of sphingofungin F.[7]

## 10.7 Shiozaki's total synthesis of sphingofungin E (2001)

See (Scheme 140).

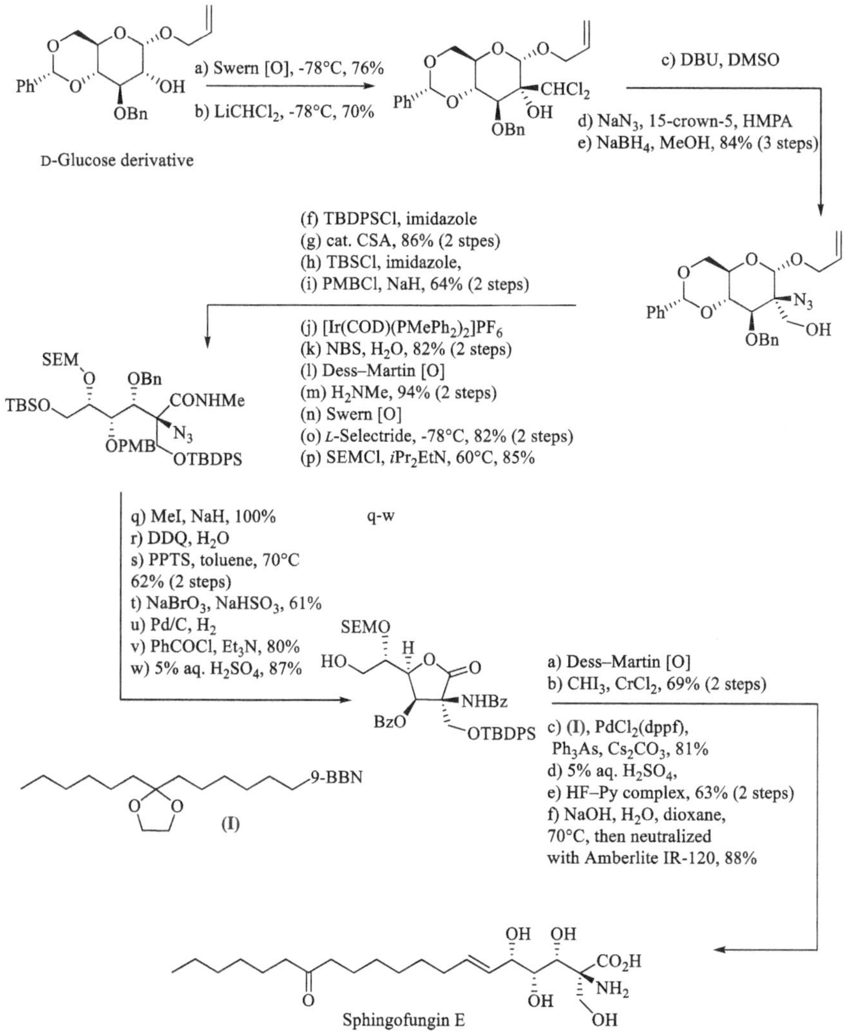


Total synthesis was completed with 1.1% overall yield in 29 steps from D-glucose derivative

SCHEME 140 Total synthesis of sphingofungin E.[8]

## 10.8 Lin's total synthesis of sphingofungin E (2001)

See (Scheme 141).

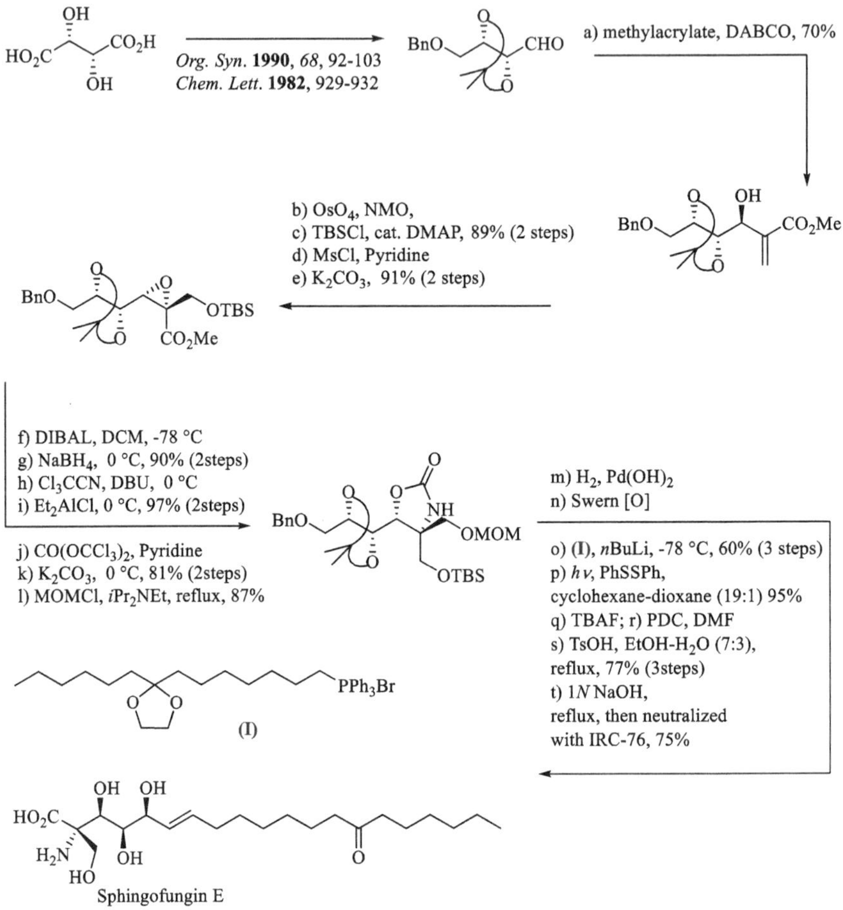


Total synthesis was completed with 8.1% overall yield in 19 steps from L-(+)-tartaric acid

SCHEME 141 Total synthesis of sphingofungin E.[9]

## 10.9 Chida's total synthesis of sphingofungin E (2002)

See (Scheme 142).

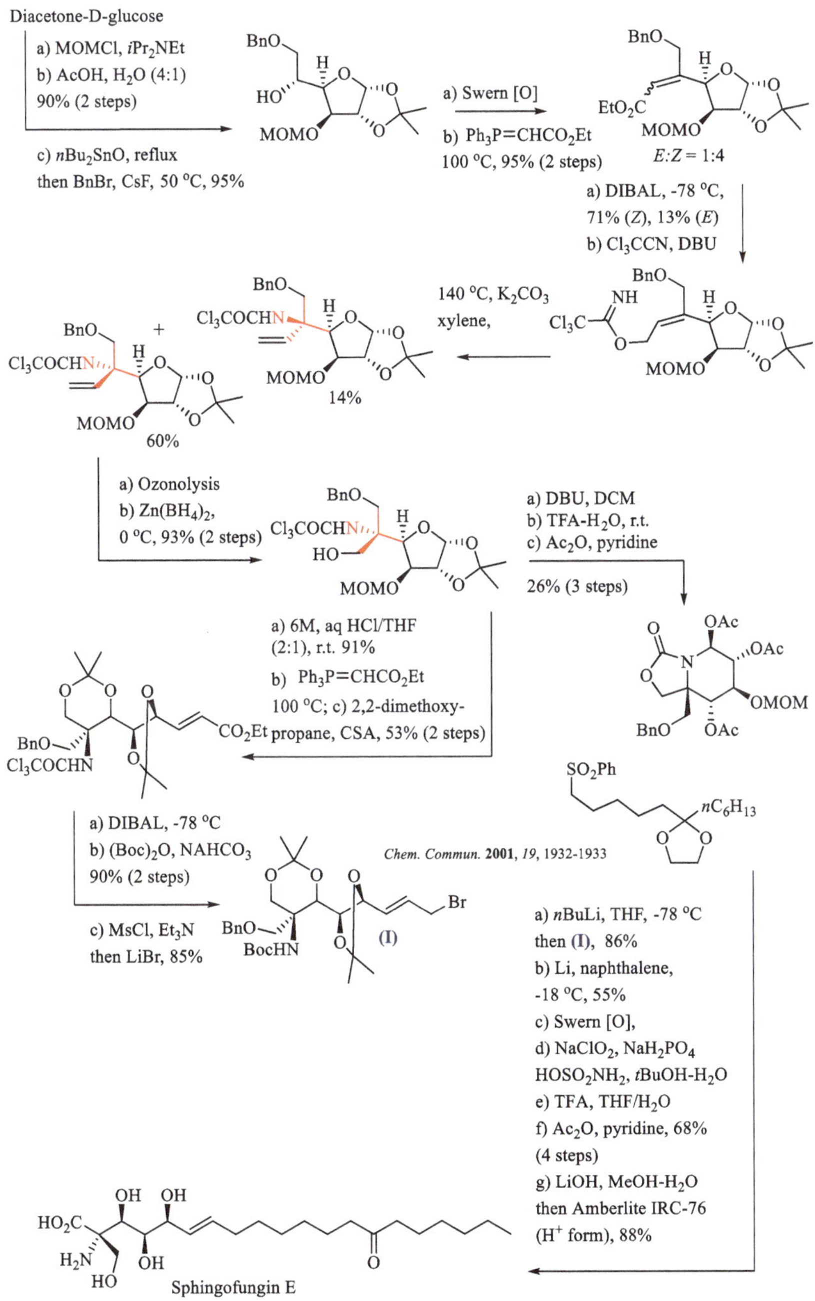


**SCHEME 142** Total synthesis of sphingofungin E from D-Glucose.[10]

## 10.10 Chida's total synthesis of sphingofungin E (2002)

See (Scheme 143).

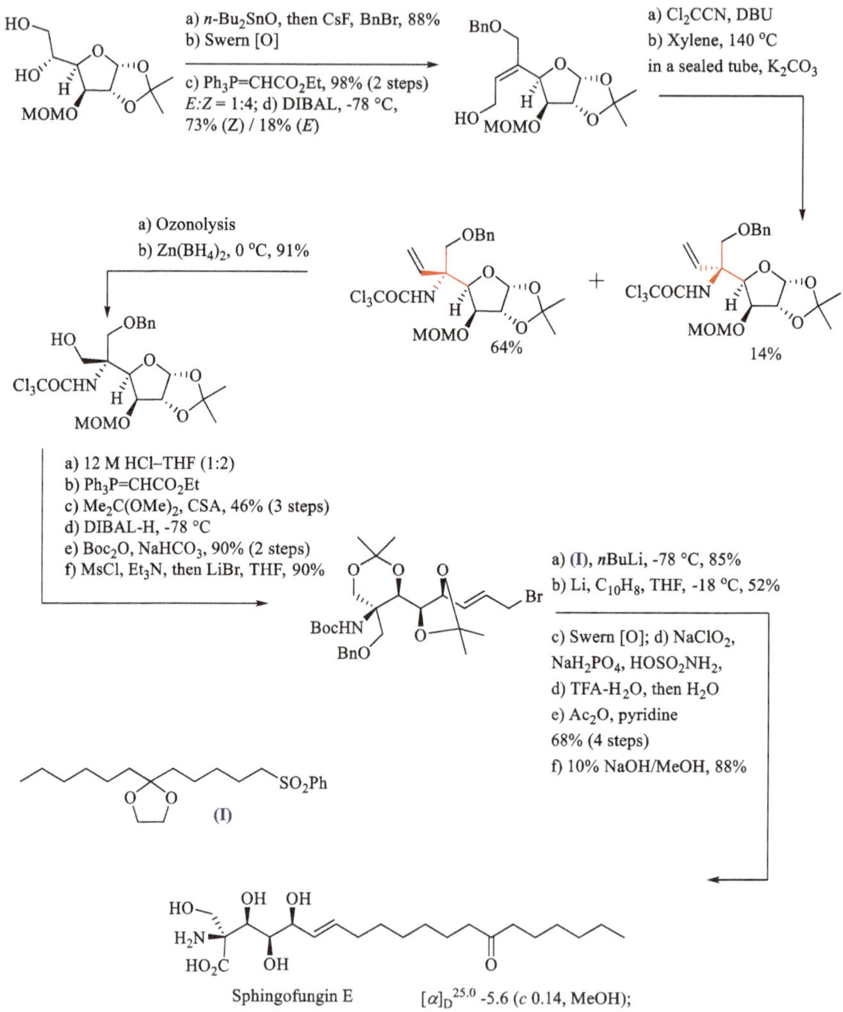


SCHEME 143 Total synthesis of sphingofungin E via Overman rearrangement.[11]

## 10.11 Ham's total synthesis of sphingofungin F (2002)

See (Scheme 144).

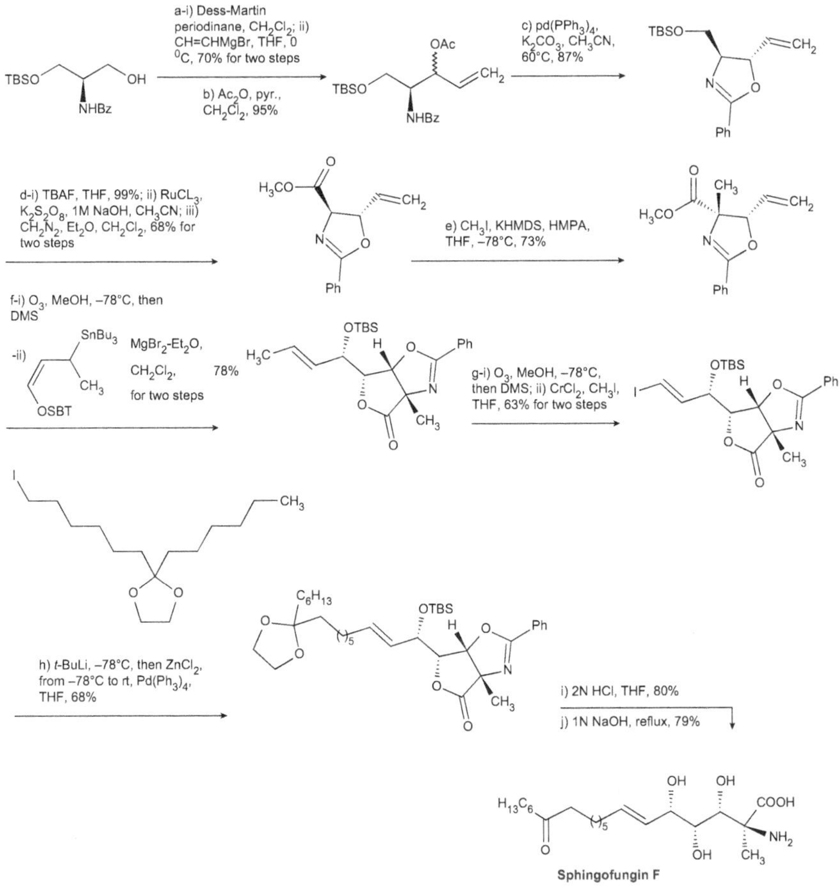


SCHEME 144 Total synthesis of sphingofungin F.[12]

## 10.12 Hayes's approach toward sphingofungin E (2006)[13]

See (Scheme 145).

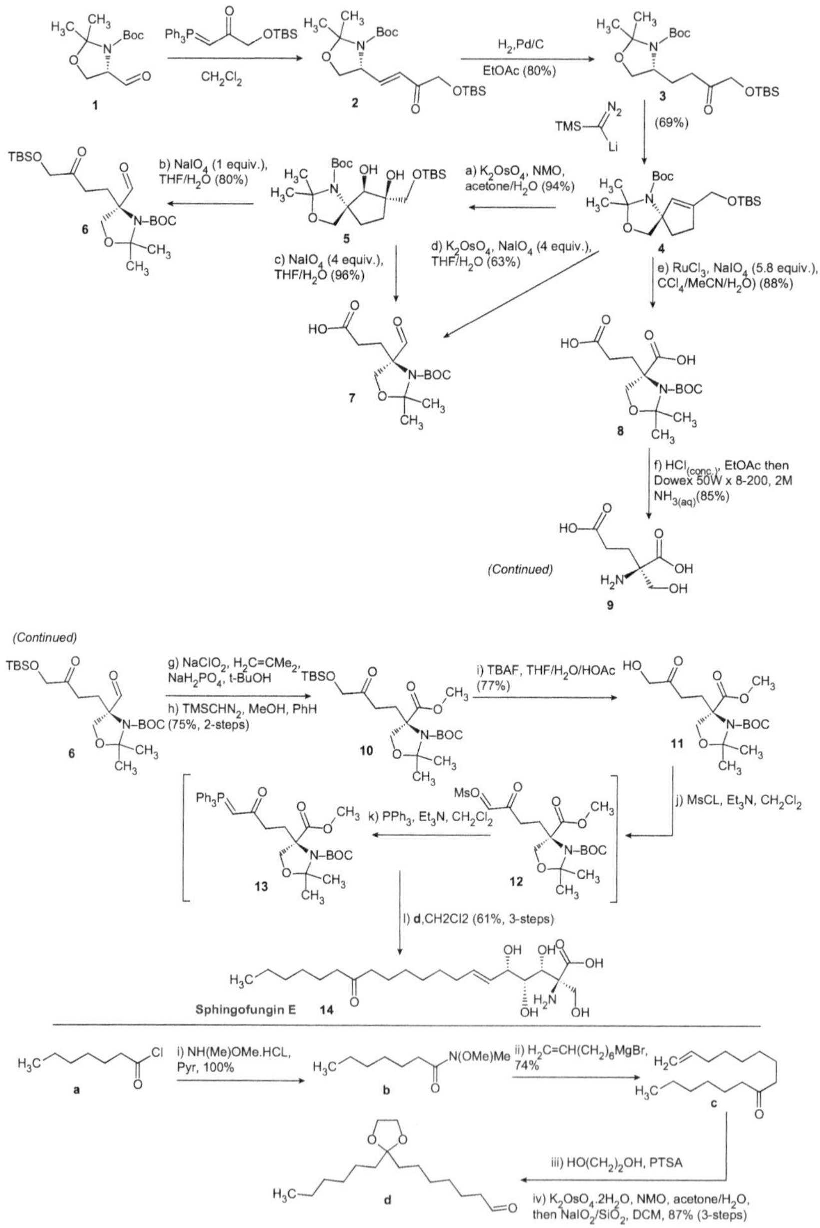


**SCHEME 145** Synthetic studies toward sphingofungin E.

## 10.13 Xu's total synthesis of sphingofungin F (2010)

See (Scheme 146).

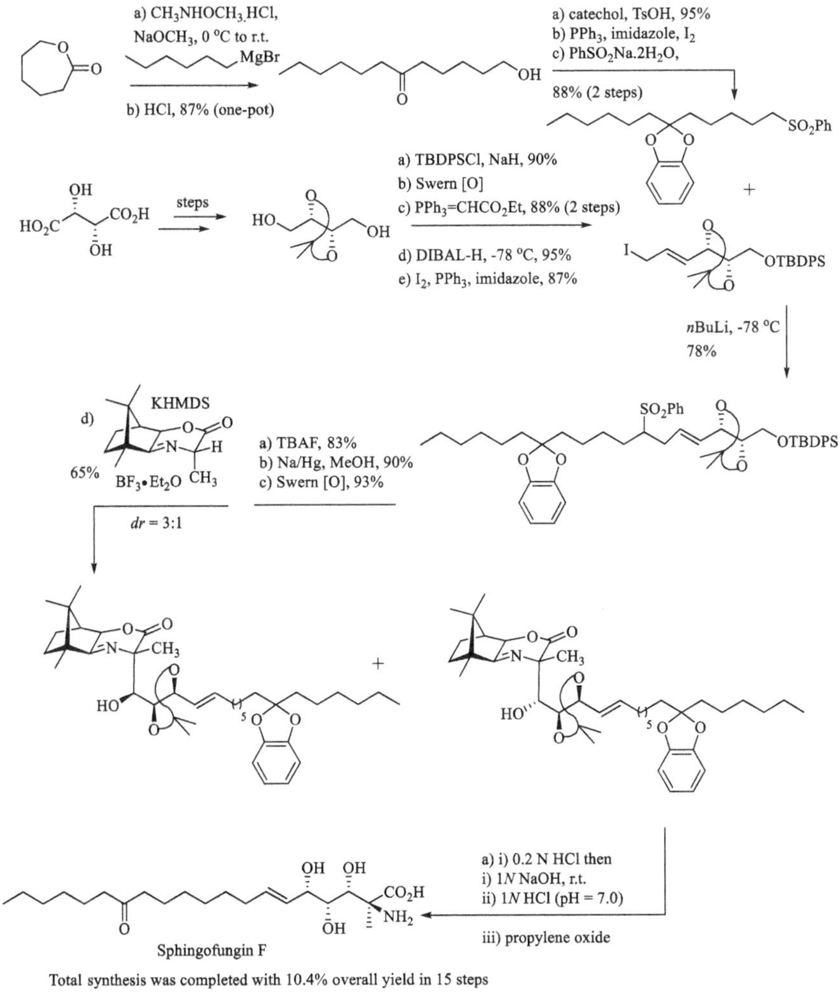


**SCHEME 146** Total synthesis of sphingofungin F.[14]

## 10.14 Martinková's total synthesis of sphingofungin E (2010)

See (Scheme 147).

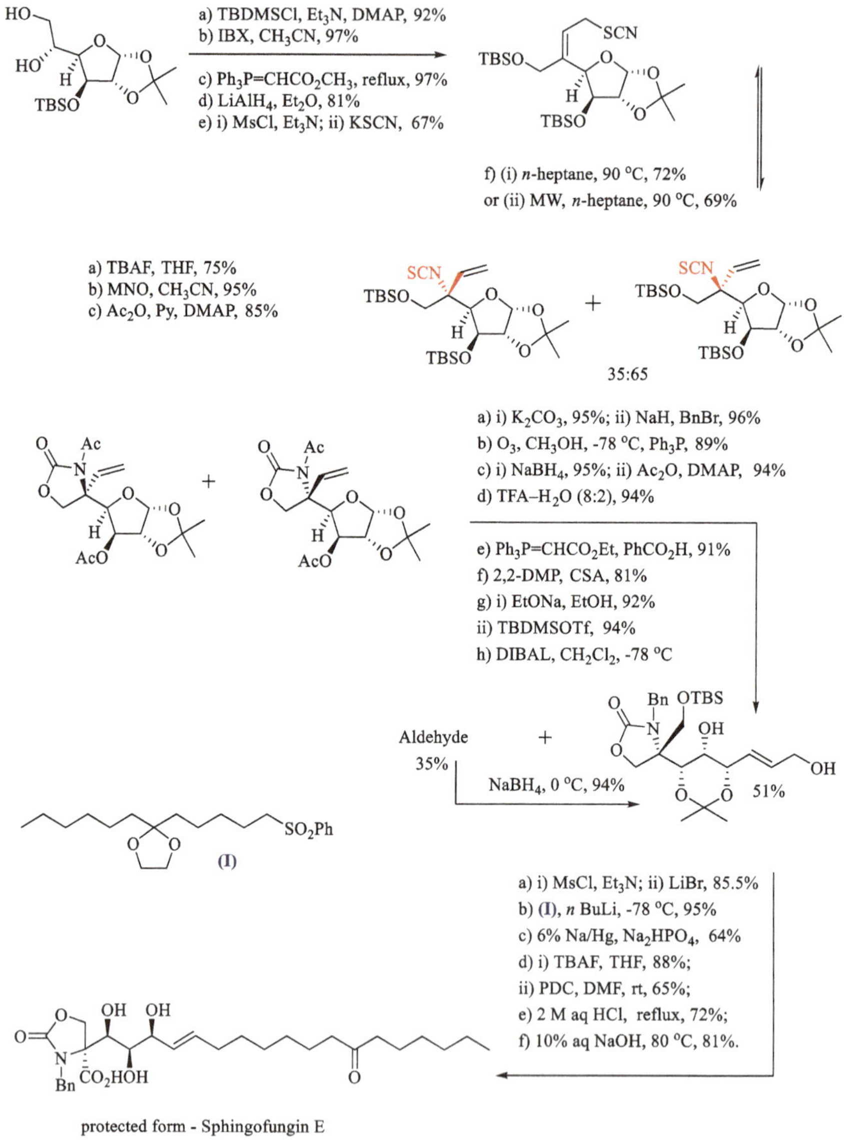


SCHEME 147 Total synthesis of a protected form of sphingofungin E.[15]

## 10.15 Kan's total synthesis of sphingofungin E (2013)

See (Scheme 148).

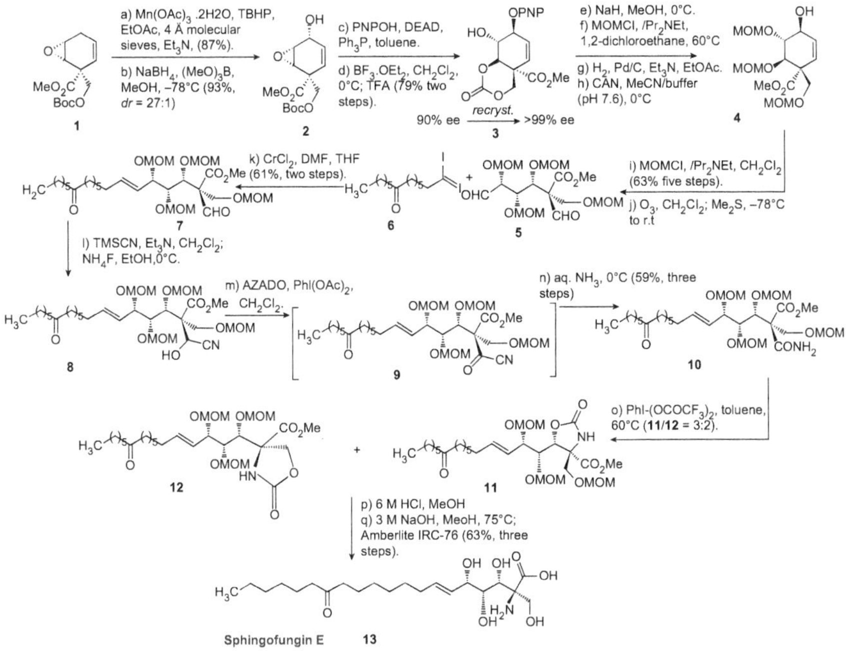


**SCHEME 148** Total synthesis of sphingofungin E.[16]

## 10.16 Chida's total synthesis of sphingofungin F (2015)

See (Scheme 149).

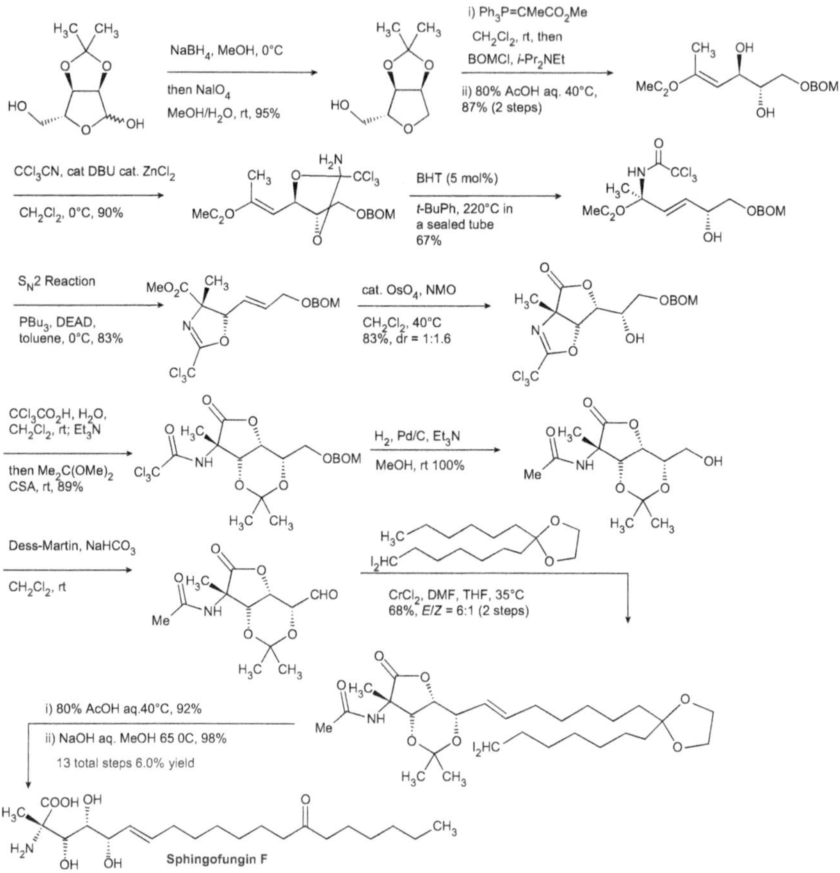


**SCHEME 149** Total synthesis of sphingofungin F by orthoamide-type Overman rearrangement.[17]

## 10.17 Yakura's total synthesis of sphingofungin E (2017)

See (Scheme 150).

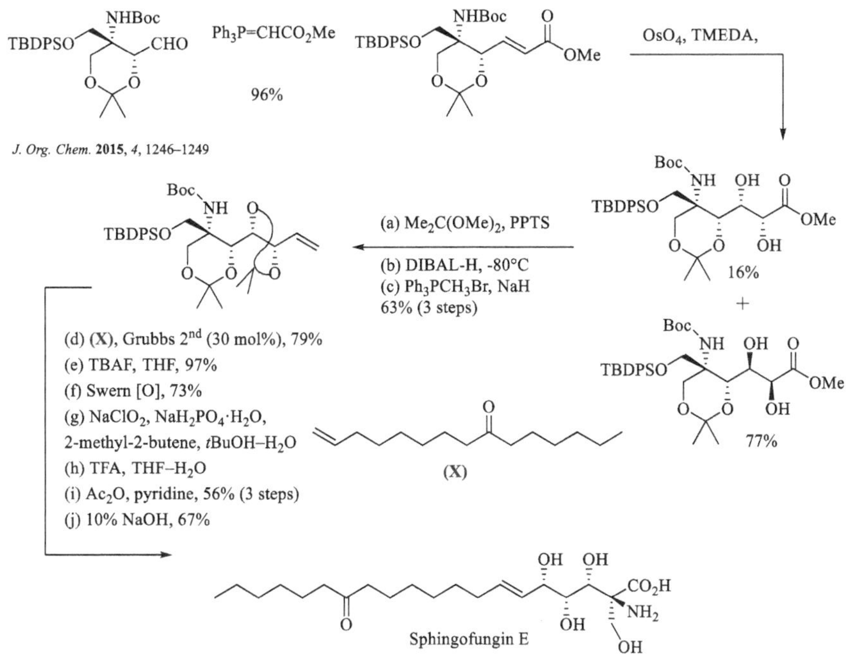


**SCHEME 150** Total synthesis of sphingofungin E.[18]

## References


1. Horn, W. S.; Smith, J. L.; Bills, G. F.; Raghoobar, S. L.; Helms, G. L.; Kurtz, M. B.; Marrinan, J. A.; Frommer, B. R.; Thornton, R. A.; Mandala, S. M. Sphingofungins E and F: Novel Serinepalmitoyl Trans-ferase Inhibitors from *Paecilomyces Variotii*. *J. Antibiot.* **1992,** *45,* 1692–1696.
2. Zweerink, M. M.; Edison, A.; Wells, G.; Pinto, W.; Lester, R. Characterization of a Novel, Potent, and Specific Inhibitor of Serine Palmitoyltransferase. *J. Biol. Chem.* **1992,** *267,* 25032–25038.
3. Delgado, A.; Casas, J.; Llebaria, A.; Abad, J. L.; Fabrias, G. Inhibitors of Sphingolipid Metabolism Enzymes. *Biochim. Biophys. Acta* **2006,** *1758,* 1957–1977.
4. Kobayashi, S.; Furuta, T. Use of Heterocycles as Chiral Ligands and Auxiliaries in Asymmetric Syntheses of Sphingosine, Sphingofungins B and F. *Tetrahedron* **1998,** *54,* 10275–10294.
5. Trost, B. M.; Lee, C. B. A New Strategy for the Synthesis of Sphingosine Analogues. Sphingofungin F. *J. Am. Chem. Soc.* **1998,** *120,* 6818–6819.
6. Trost, B. M.; Lee, C. *gem*-Diacetates as Carbonyl Surrogates for Asymmetric Synthesis. Total Syntheses of Sphingofungins E and F. *J. Am. Chem. Soc.* **2001,** *123,* 12191–12201.

7. Liu, D.-G.; Wang, B.; Lin, G.-Q. An Efficient and Convenient Approach to the Total Synthesis of Sphingofungin. *J. Org. Chem.* **2000,** *65,* 9114–9119.
8. Nakamura, T.; Shiozaki, M. Total Synthesis of Sphingofungin E. *Tetrahedron Lett.* **2001,** *42,* 2701–2704.
9. Wang, B.; Yu, X.-m.; Lin, G.-q. The First Total Synthesis of Sphingofungin E and the Determination of Its Stereochemistry. *Synlett* **2001,** *2001* (Special Issue), 0904–0906.
10. Oishi, T.; Ando, K.; Inomiya, K.; Sato, H.; Iida, M.; Chida, N. Total Synthesis of Sphingofungin E From D-glucose. *Org. Lett.* **2002,** *4,* 151–154.
11. Oishi, T.; Ando, K.; Inomiya, K.; Sato, H.; Iida, M.; Chida, N. Total Synthesis of (+)-myriocin and (−)-sphingofungin E From Aldohexoses Using Overman Rearrangement as the Key Reaction. *Bull. Chem. Soc. Jpn.* **2002,** *75,* 1927–1947.
12. Lee, K.-Y.; Oh, C.-Y.; Ham, W.-H. Total Synthesis of Sphingofungin F. *Org. Lett.* **2002,** *4,* 4403–4405.
13. Hayes, C. J.; Bradley, D. M.; Thomson, N. M. An Efficient Enantioselective Synthesis of (2*R*)-Hydroxymethyl Glutamic Acid and an Approach to the (2*R*)-Hydroxymethyl-Substituted Sphingofungins. *J. Org. Chem.* **2006,** *71* (7), 2661–2665.
14. Gan, F.-F.; Yang, S.-B.; Luo, Y.-C.; Yang, W.-B.; Xu, P.-F. Total Synthesis of Sphingofungin F Based on Chiral Tricyclic Iminolactone. *J. Org. Chem.* **2010,** *75,* 2737–2740.
15. Martinková, M.; Gonda, J.; Raschmanová, J.Š.; Slaninková, M.; Kuchár, J. Total Synthesis of a Protected Form of Sphingofungin E Using the [3, 3]-sigmatropic Rearrangement of an Allylic Thiocyanate as the Key Reaction. *Carbohydrate Res.* **2010,** *345,* 2427–2437.
16. Ikeuchi, K.; Hayashi, M.; Yamamoto, T.; Inai, M.; Asakawa, T.; Hamashima, Y.; Kan, T. Stereocontrolled Total Synthesis of Sphingofungin E. *Europ. J. Org. Chem.* **2013,** *2013,* 6789–6792.
17. Tsuzaki, S.; Usui, S.; Oishi, H.; Yasushima, D.; Fukuyasu, T.; Oishi, T.; Sato, T.; Chida, N. Total Synthesis of Sphingofungin F by Orthoamide-type Overman Rearrangement of an Unsaturated Ester. *Org. Lett.* **2015,** *17,* 1704–1707.
18. Noda, N.; Nambu, H.; Fujiwara, T.; Yakura, T. Total Synthesis of Sphingofungin E and 4, 5-Di-epi-sphingofungin E. *Chem. Pharm. Bull.* **2017,** *65* (7), 687–696.

CHAPTER

# 11

# (−)-FR901483 and TAN1251 (A-D)

## 11.1 Abstract

The immunosuppressant (−)-FR901483 was isolated in 1996 by scientists at the Fujisawa Pharmaceutical Company in 1996 from the fermentation broth of *Cladobotryum* sp. No. 11231 collect at Iwaki, Japan.[1] The structural framework and relative stereochemistry of FR901483 was established via X-ray crystallographic data; however, the absolute configuration of the natural product was demonstrated when Sinder and Lin[154] reported the first total synthesis of this molecule. The structural framework of natural product showed that it exhibits a core tricyclic ring that contains 2-azabicyclo[3.3.1]nonane core fused to a pyrrolidine moiety as well as phosphate residue necessary for bioactivity. The natural was found to have promising immunosuppressant activity in vitro and showed significant prolong graft survival time in the rat skin allograft model. Further, the biological studies of FR901483 showed nonspecific immunosuppressive profile that suggests natural product role as antimetabolite by inhibiting adenylosuccinate synthetase and/or adenylosuccinate lyase. The enzymes have important role in de novo adenosine biosynthetic pathway.[2] Moreover, the different modes of action FR901483 from the existing immunosuppressants, i.e., cyclosporine A and tacrolimus suggest it an effective therapeutic agent for new drug development (Fig. 11).

The structures of TAN1251 alkaloids also exhibit tricyclic ring core that consists of 1,4-diazabicyclo[3.2.1]octane ring and a spiro-fused cyclohexanone ring and side chain isoprenyl unit. This class of natural products was isolated from *Penicillium thomii* RA-89 at Takeda Chemical Industries Ltd. in Japan.[3] The biological profile of these natural products


https://doi.org/10.1016/B978-0-12-822262-1.00011-1

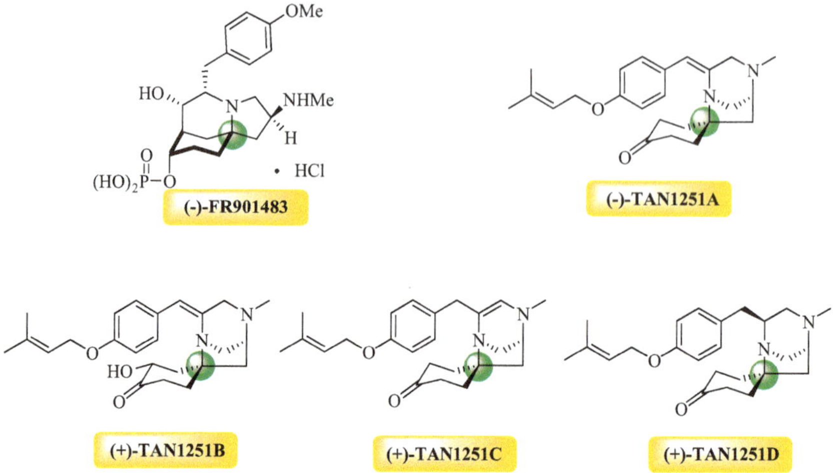


FIGURE 11 Structures of FR901483 and TAN1251A-D.

showed TAN1251A and TAN1251B as mydriatic or antispasmodic/anti-ulcer agents. Moreover, the affinity of TAN1251B showed stronger affinity for muscarinic acetylcholine receptor than atropine, an anticholinergic or antiparasympathetic drug.[4]

## 11.2 Proposed biosynthesis of FR901483 and TAN1251

See (Scheme 151).

## 11.3 Sinder's total synthesis of (−)-FR901483 (1999)

See (Scheme 152).

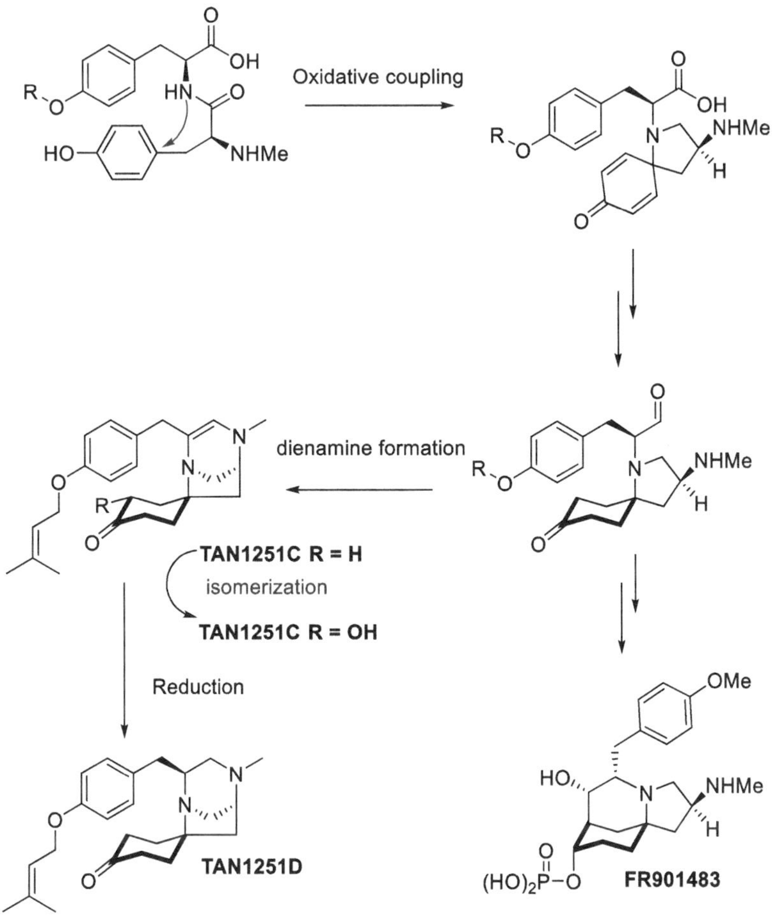


SCHEME 151 Proposed biosynthesis of FR901483 and TAN1251.[5]

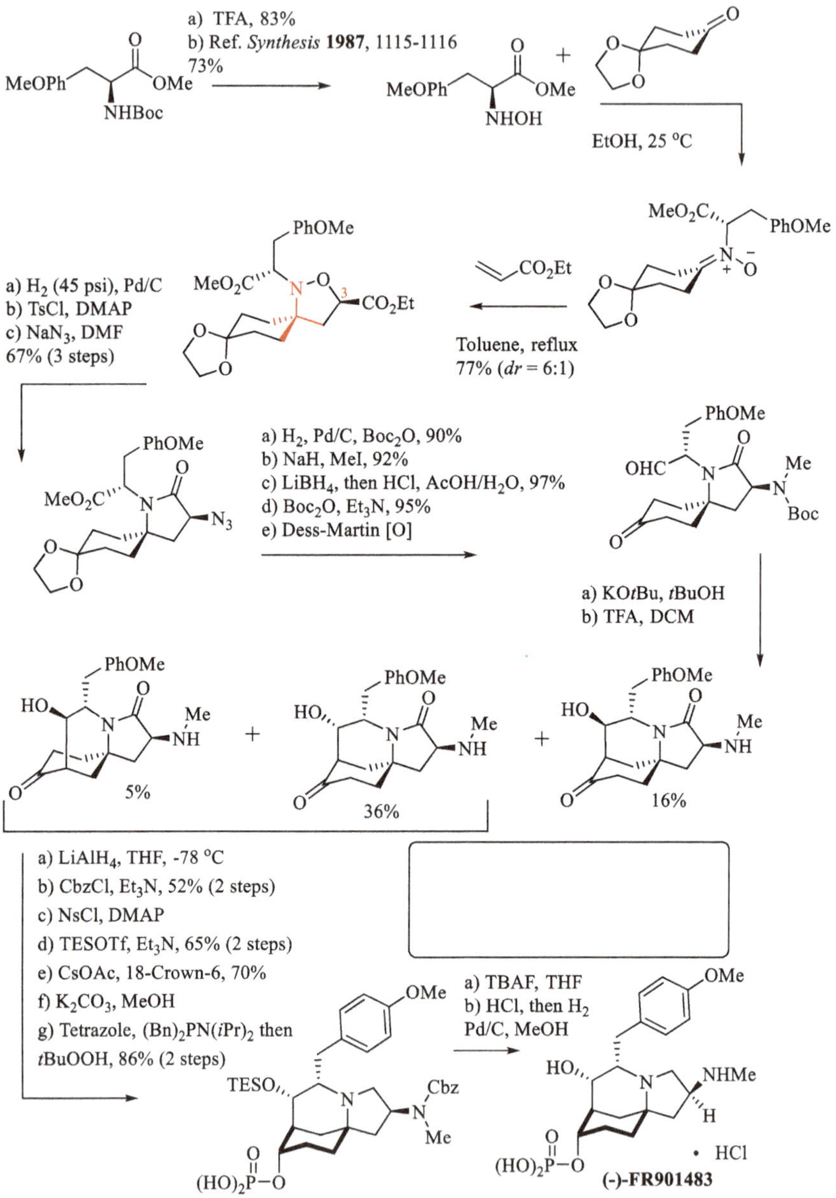


Total Synthesis 2% overall yield from *O*-methyltyrosine methyl ester in 22 steps
$[\alpha]_D$ -10 (*c* 0.37, MeOH); Lit. $[\alpha]_D$ -11 (*c* 0.74, MeOH) *J. Antibiot.* **1996**, *49*, 37-44.

SCHEME 152 Total synthesis of (−)-FR901483.[6]

## 11.4 Sorensen's synthesis via oxidative cyclization (2000)

See (Scheme 153).

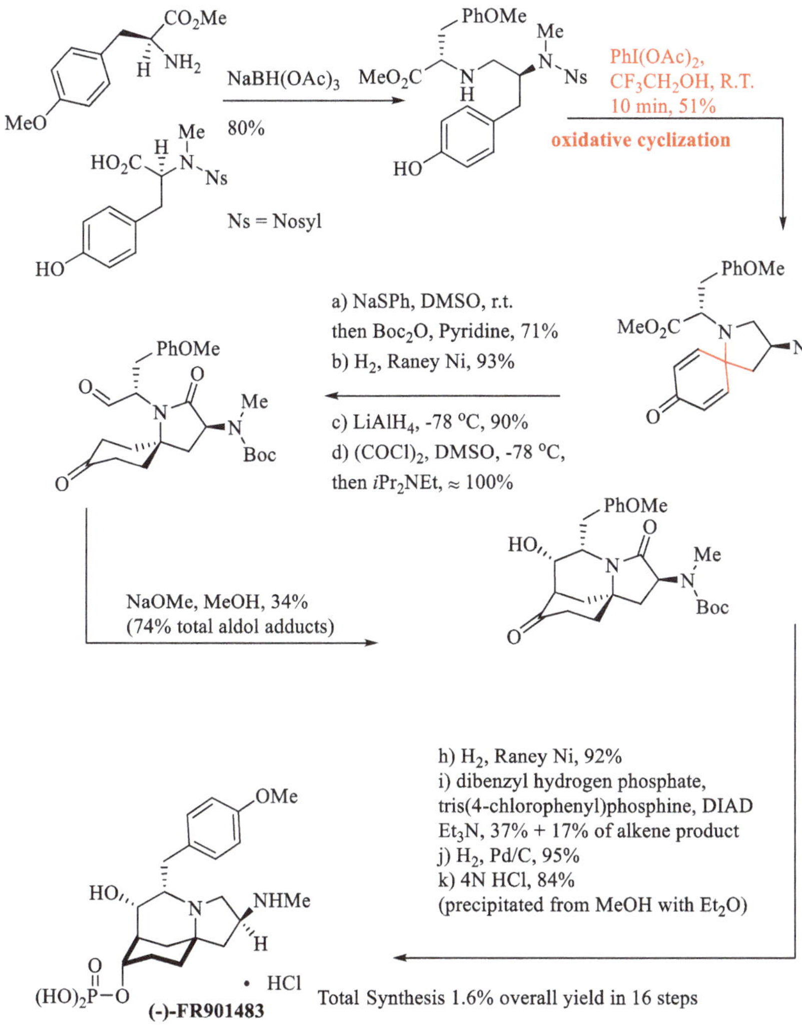


SCHEME 153 FR901483 synthesis via oxidative cyclization.[7]

## 11.5 Ciufolini's synthesis via oxidative cyclization (2001)

See (Scheme 154).

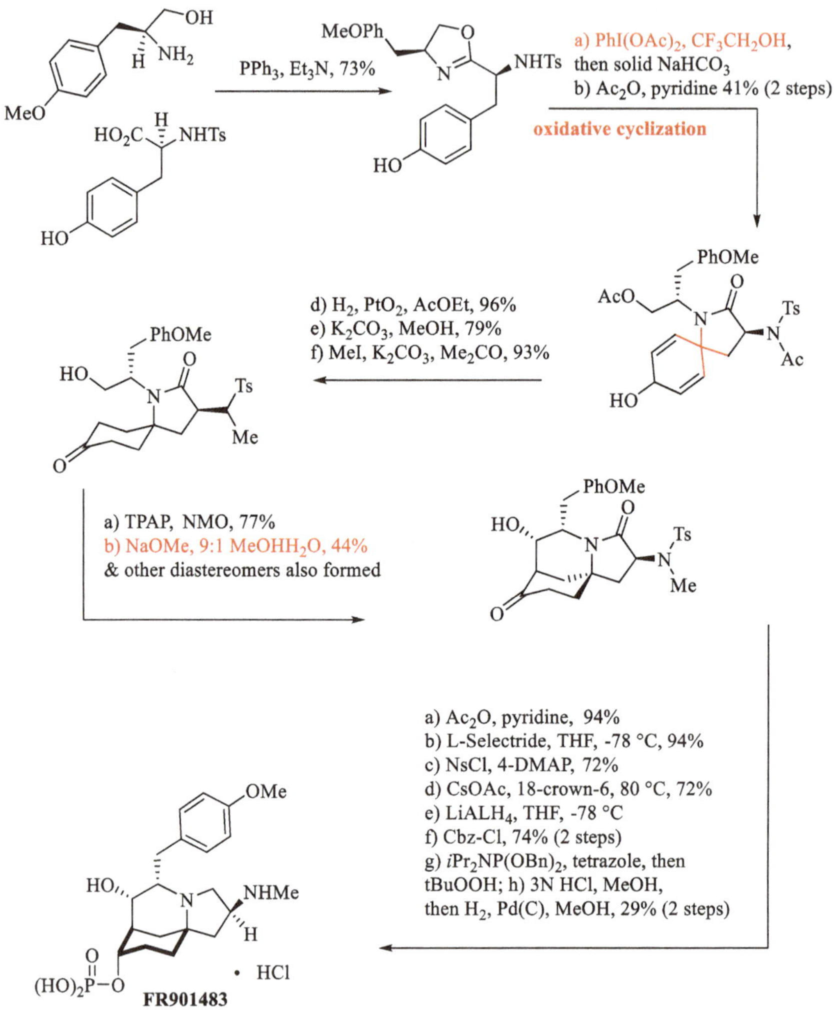


**FR901483 *bis*-hydrochloride**, $[\alpha]_D^{25}$ +4.0° (*c* 0.35, $CH_3OH$)

Lit. +5° (*J. Am. Chem. Soc.* **1999**, *121*, 7778)

SCHEME 154 Synthesis via oxidative cyclization[8,9].

## 11.6 Funk's total synthesis (2001)

See (Scheme 155).

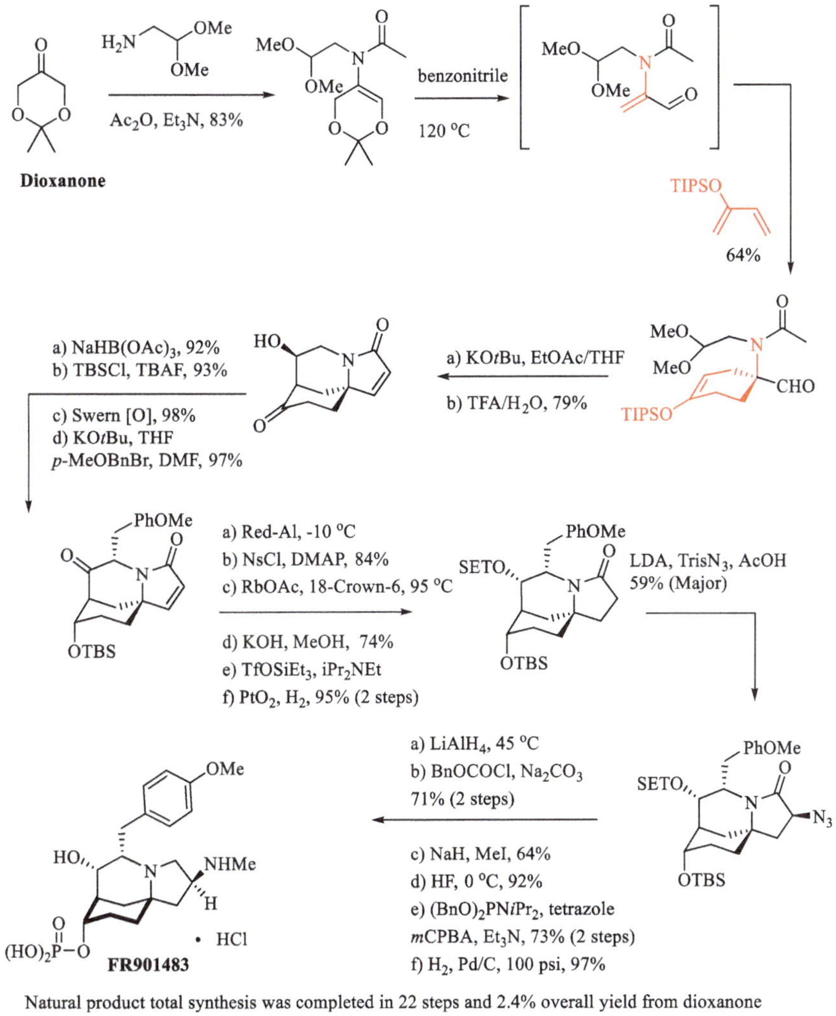


**SCHEME 155** Total synthesis via an amidoacrolein cycloaddition.[10]

## 11.7 Wardrop's formal synthesis of (±)-desmethylamino FR901483 (2001)

See (Scheme 156).

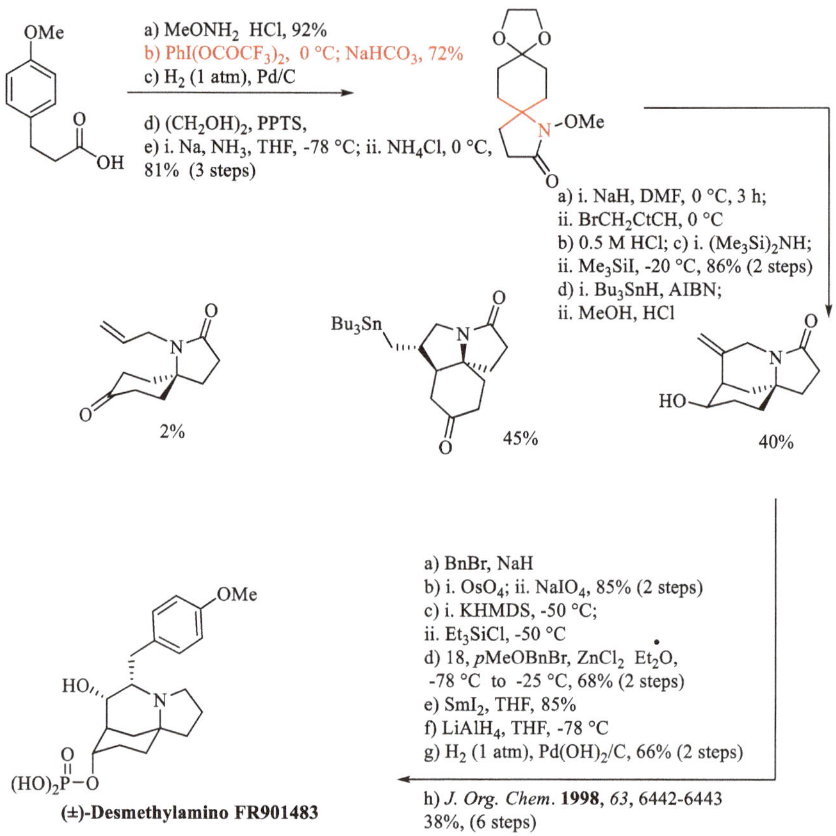


SCHEME 156 Formal Synthesis of (±)-Desmethylamino FR901483 Via an *N*-Methoxy-*N*-Acylnitrenium Ion-Induced Spirocyclization.[11]

## 11.8 Fukuyama's total synthesis (2004)

See (Scheme 157).

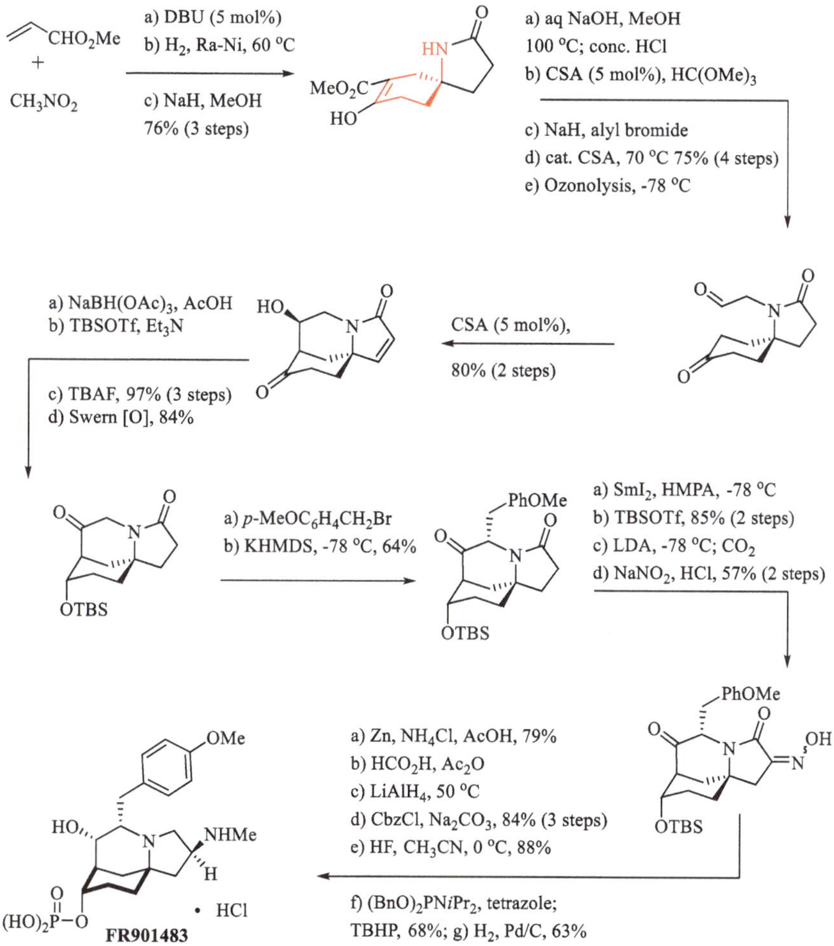


SCHEME 157 Stereocontrolled total synthesis.[12]

## 11.9 Brummond's formal synthesis (2005)

See (Scheme 158).

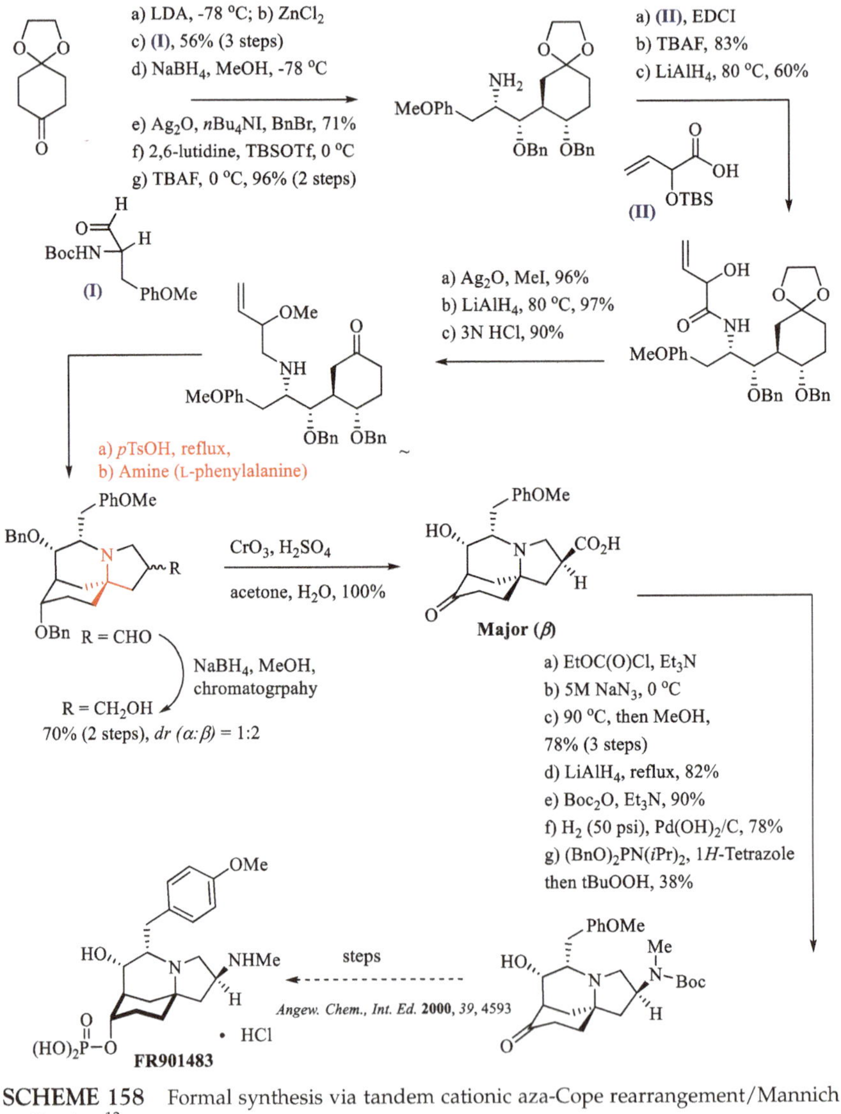


SCHEME 158 Formal synthesis via tandem cationic aza-Cope rearrangement/Mannich cyclization.[13]

## 11.10 Kerr's total synthesis via ring-opening/annulation reaction (2009)

See (Scheme 159).

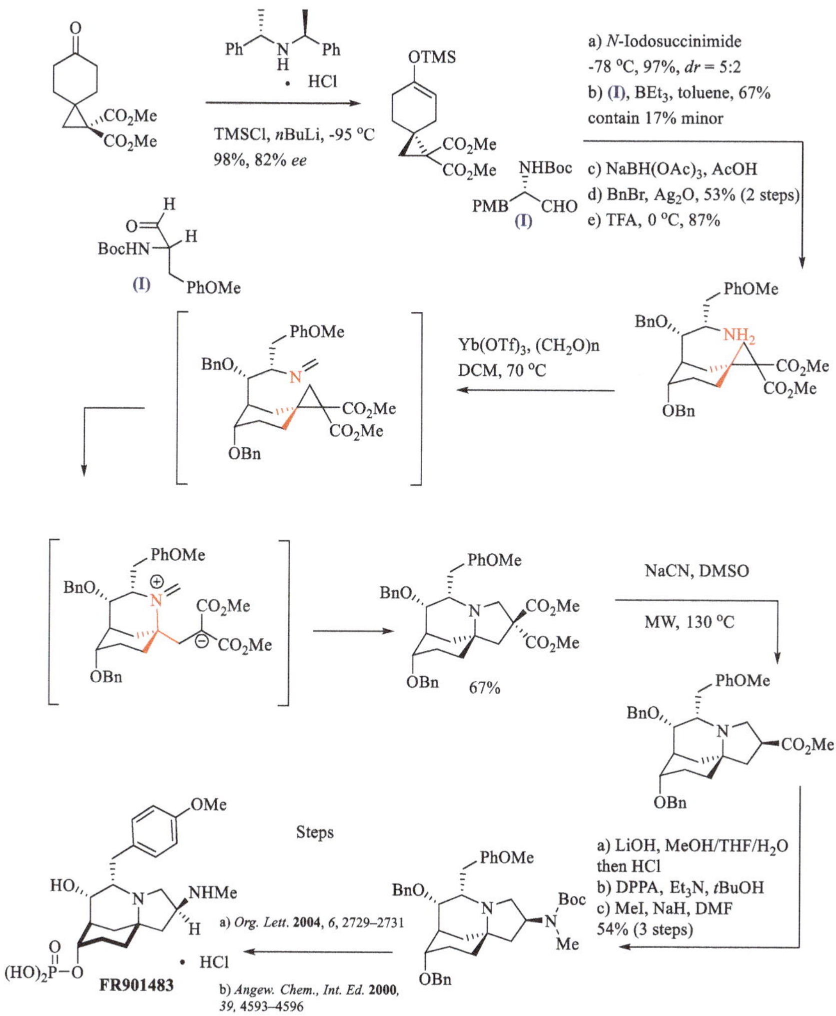


**SCHEME 159** Total synthesis via ring-opening/annulation reaction.[14]

## 11.11 Fukuyama's intermediate synthesis of FR901483 (2010)

See (Scheme 160).

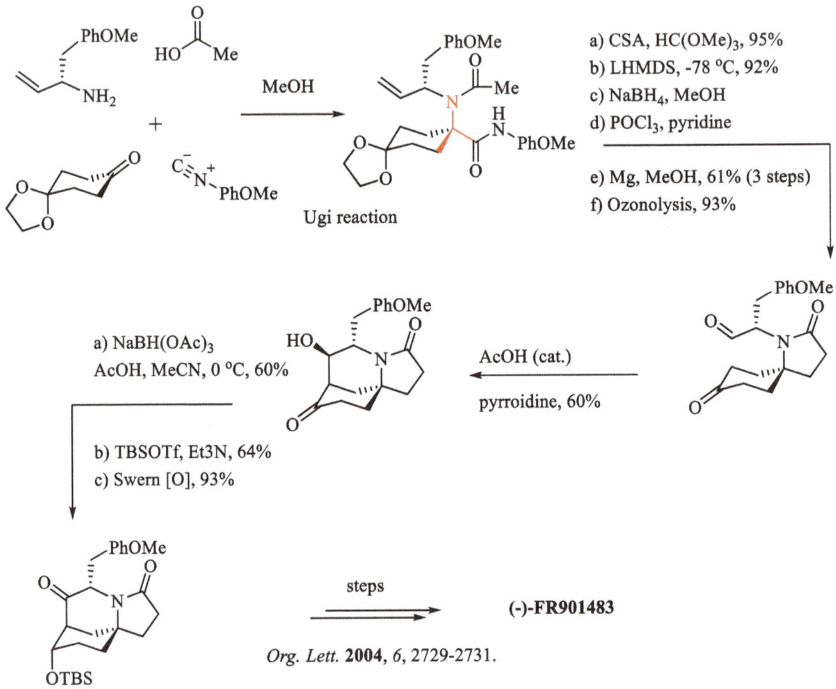


SCHEME 160 Optically active key intermediate of FR901483.[15]

## 11.12 Bonjoch's tricyclic skeleton of FR901483 (2003)

See (Scheme 161).

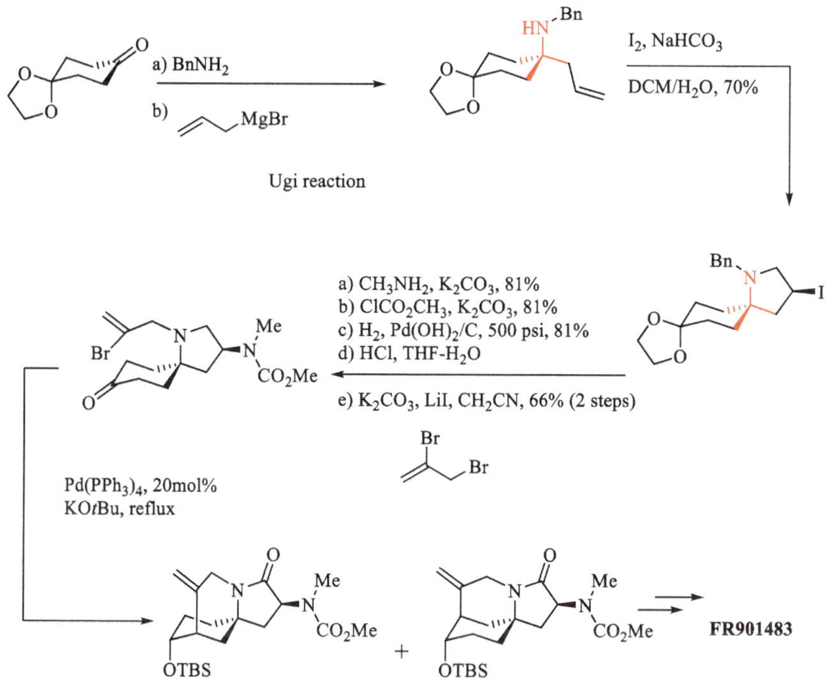


SCHEME 161 Tricyclic skeleton of FR901483 By palladium-catalyzed cyclization.[16]

## 11.13 Weinreb's studies toward total synthesis (2006)

See (Scheme 162).

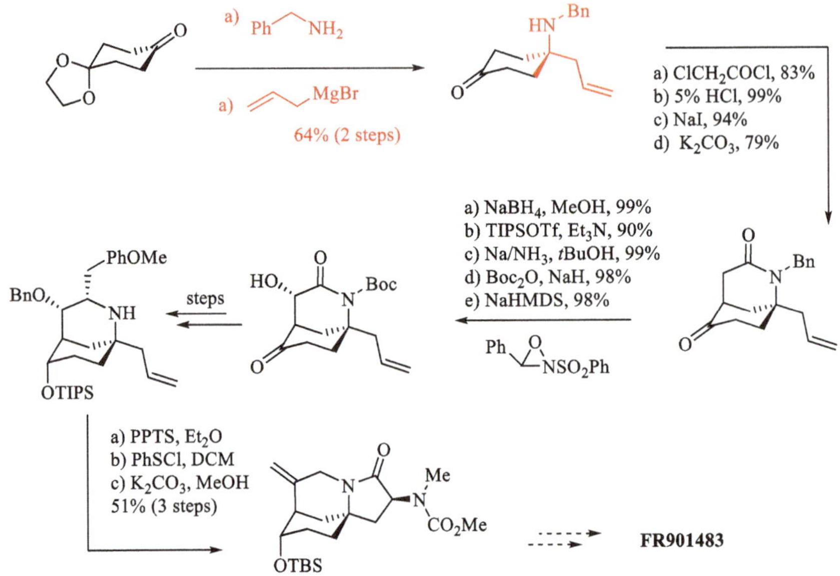


SCHEME 162 Studies toward total synthesis.[17]

## 11.14 Reissig's approach toward azaspirane core of FR901483 (2006)

See (Scheme 163).

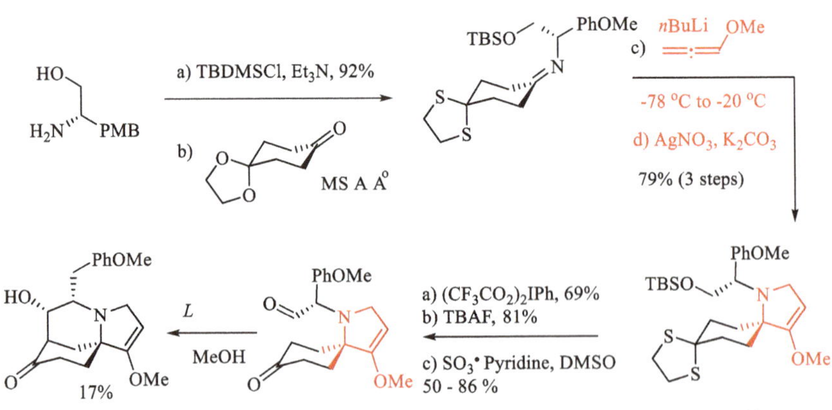


SCHEME 163 Synthetic studies toward azaspirane core of FR901483.[18]

# 11.15 Huang's formal enantioselective synthesis of (−)-FR901483 (2012)

See (Scheme 164).

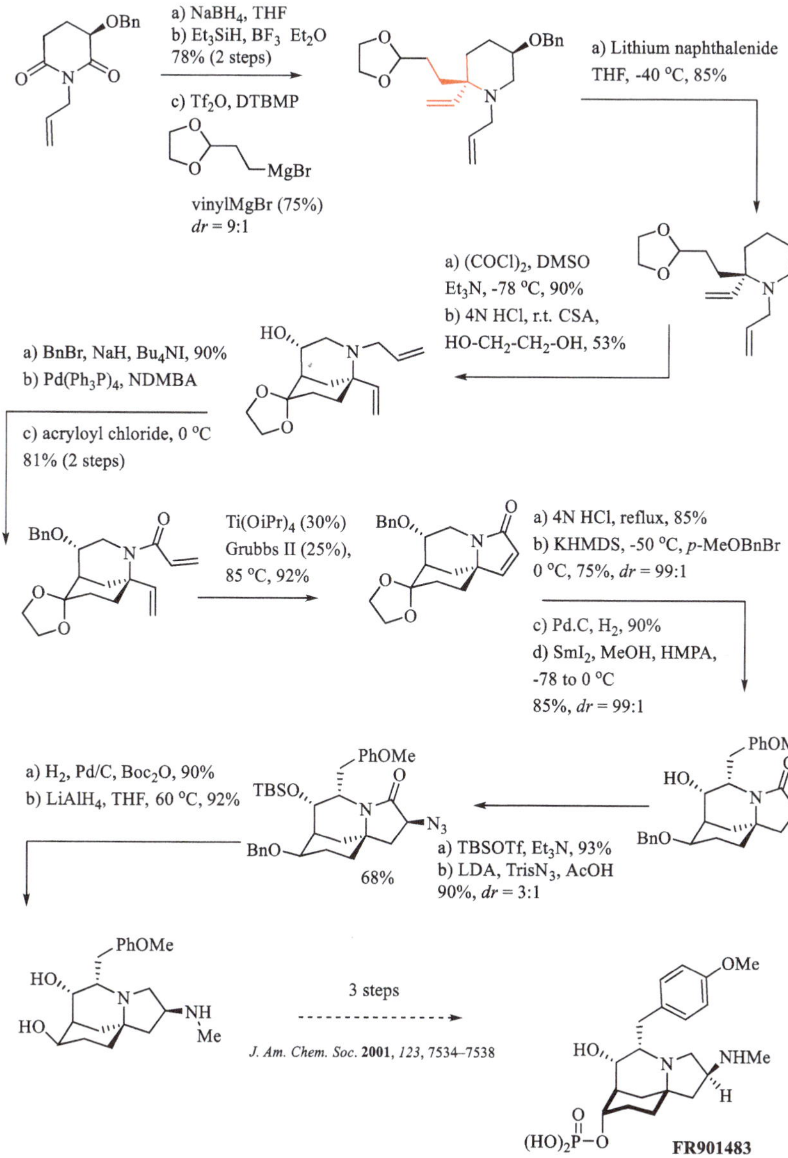


SCHEME 164 Formal enantioselective synthesis of (−)-FR901483 via one-pot amide reductive bisalkylation method.[19]

## 11.16 Huang's enantioselective total syntheses of (−)-FR901483 and (+)-8-*epi*-FR901483 (2013)

See (Scheme 165).

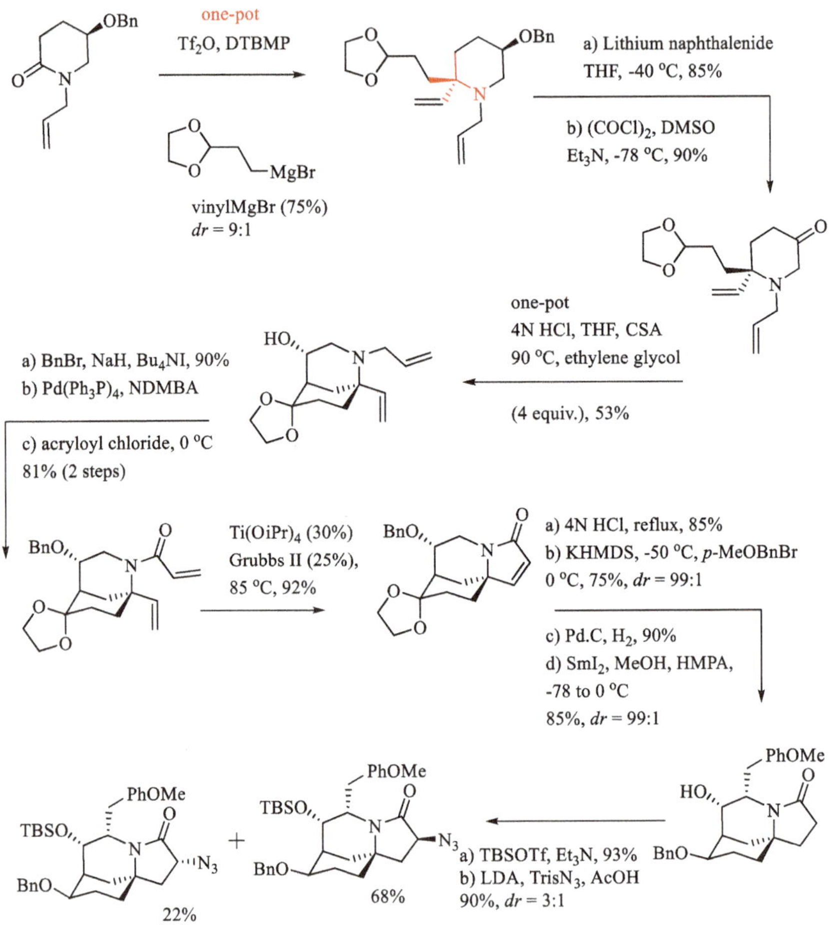

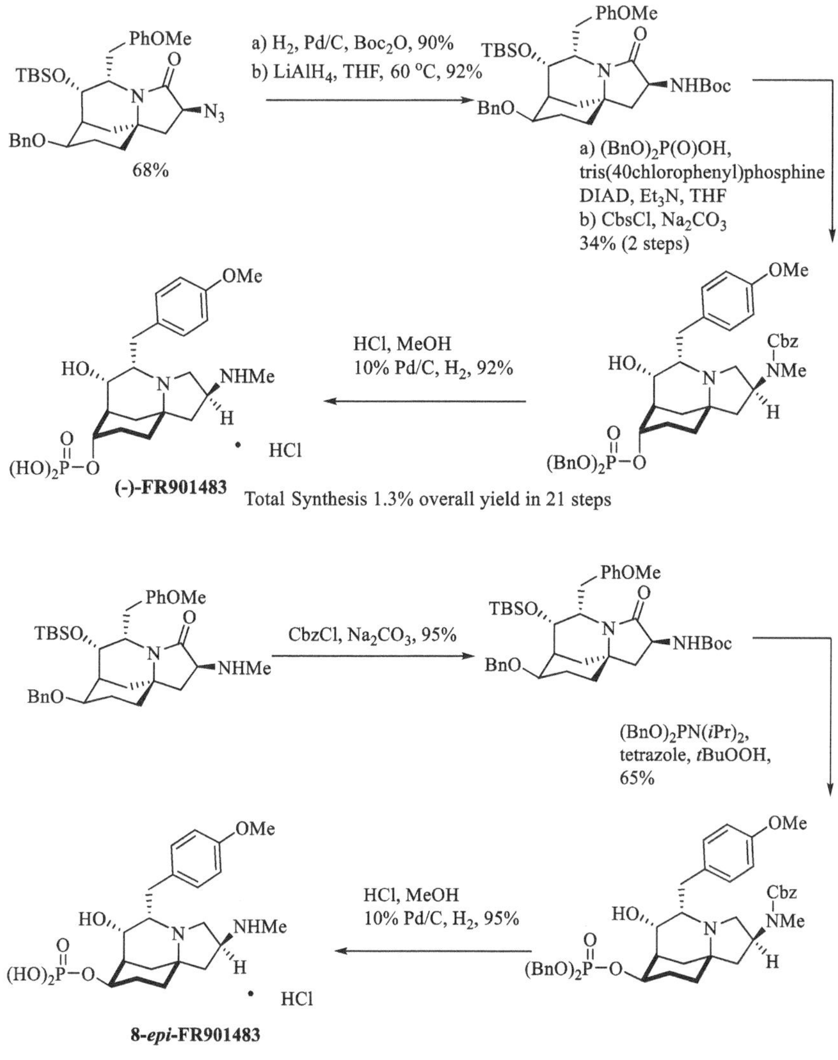


SCHEME 165 Enantioselective total syntheses of (−)-FR901483 and (+)-8-*epi*-FR901483.[20]

## 11.17 Gaunt's syntheses of (−)-FR901483 and (+)-TAN1251C (2019)

See (Scheme 166).

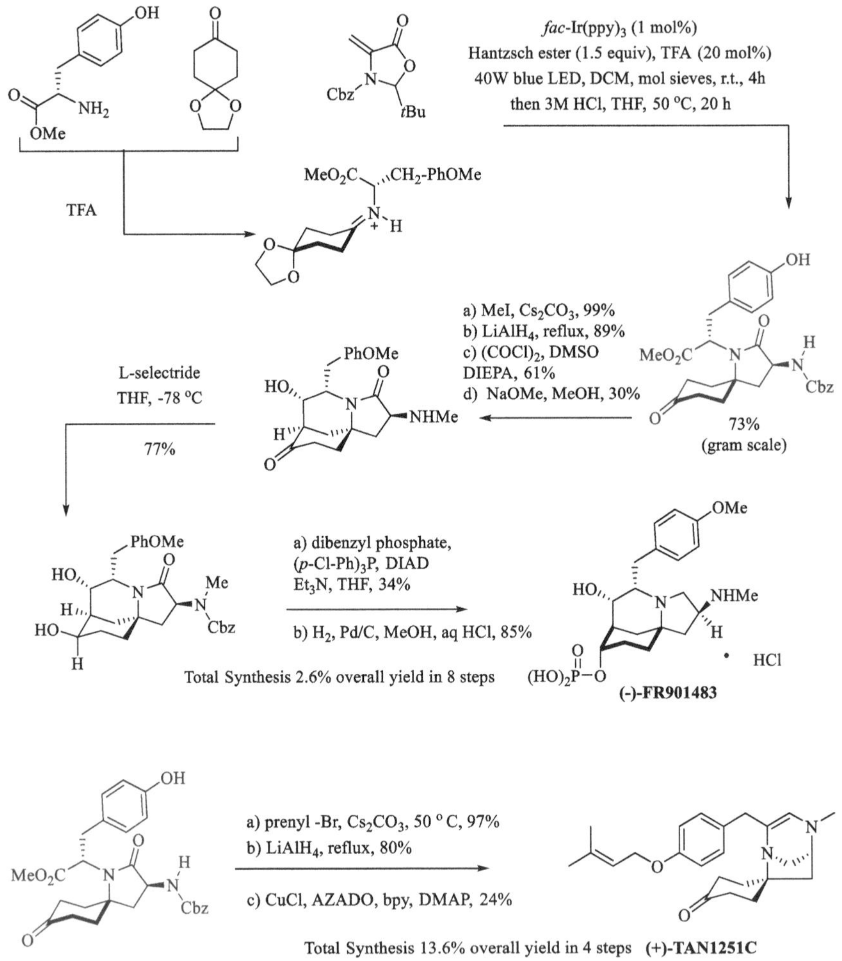


SCHEME 166 Syntheses of (−)-FR901483 and (+)-TAN1251C via photocatalytic olefin hydroaminoalkylation.[21]

# References


1. Sakamoto, K.; Tsujii, E.; Abe, F.; Nakanishi, T.; Yamashita, M.; Shigematsu, N.; Izumi, S.; Okuhara, M. FR901483, a Novel Immunosuppressant Isolated from *Cladobotryum* sp. No. 11231. *J. Antibiot.* **1996,** *49* (1), 37–44.
2. Ruan, Z.; Li, C.; Shen, D.; Huang, S.-H.; Hong, R. FR901483: Synthetic Efficiency Remains a Challenge. *Synthesis* **2019,** *51* (11), 2237–2251.
3. Shirafuji, H.; Tsubotani, S.; Ishimaru, T.; Harada, S. *PCT Int. Appl. 1991, WO 91/13,887,* 1992.
4. Broadley, K. J.; Kelly, D. R. Muscarinic Receptor Agonists and Antagonists. *Molecules* **2001,** *6* (3), 142–193.
5. Bonjoch, J.; Diaba, F. Synthesis of Immunosuppressant FR901483 and Biogenetically Related TAN1251 Alkaloids. In *Studies in Natural Products Chemistry,* Vol. 32; Elsevier, 2005; pp 3–60.
6. Snider, B. B.; Lin, H. Total Synthesis of (−)-FR901483. *J. Am. Chem. Soc.* **1999,** *121,* 7778–7786.
7. Scheffler, G.; Seike, H.; Sorensen, E. J. An Enantiospecific Synthesis of the Potent Immunosuppressant FR901483. *Angew. Chem. Int. Ed.* **2000,** *39,* 4593–4596.
8. Ousmer, M.; Braun, N. A.; Ciufolini, M. A. Total Synthesis of FR901483. *Org. Lett.* **2001,** *Vol. 3, 765–767.*
9. Ousmer, M.; Braun, N. A.; Bavoux, C.; Perrin, M.; Ciufolini, M. A. *J. Am. Chem. Soc.* **2001,** *123,* 7534–7538.
10. Maeng, J.-H.; Funk, R. L. Total Synthesis of the Immunosuppressant FR901483 *via* an Amidoacrolein Cycloaddition. *Org. Lett.* **2001,** *3,* 1125–1128.
11. Wardrop, D. J.; Zhang, W. *N*-Methoxy-*N*-acylnitrenium Ions: Application to the Formal Synthesis of (±)-Desmethylamino FR901483. *Org. Lett.* **2001,** *3,* 2353–2356.
12. Kan, T.; Fujimoto, T.; Ieda, S.; Asoh, Y.; Kitaoka, H.; Fukuyama, T. Stereocontrolled Total Synthesis of Potent Immunosuppressant FR901483. *Org. Lett.* **2004,** *6,* 2729–2731.
13. Brummond, K. M.; Hong, S.-P. A Formal Total Synthesis of (−)-FR901483, Using a Tandem Cationic Aza-Cope Rearrangement/mannich Cyclization Approach. *J. Org. Chem.* **2005,** *70,* 907–916.
14. Carson, C. A.; Kerr, M. A. Total Synthesis of FR901483. *Org. Lett.* **2009,** *11,* 777–779.
15. Ieda, S.; Kan, T.; Fukuyama, T. Synthesis of the Optically Active Key Intermediate of FR901483. *Tetrahedron Lett.* **2010,** *51,* 4027–4029.
16. Bonjoch, J.; Diaba, F.z.; Puigbó, G.; Peidró, E.; Solé, D. A New Synthetic Entry to the Tricyclic Skeleton of FR901483 by Palladium-Catalyzed Cyclization of Vinyl Bromides with Ketone Enolates. *Tetrahedron Lett.* **2003,** *44* (46), 8387–8390.
17. Kropf, J. E.; Meigh, I. C.; Bebbington, M. W.; Weinreb, S. M. Studies on a Total Synthesis of the Microbial Immunosuppresive Agent FR901483. *J. Org. Chem.* **2006,** *71,* 2046–2055.
18. Kaden, S.; Reissig, H.-U. Efficient Approach to the Azaspirane Core of FR 901483. *Org. Lett.* **2006,** *8,* 4763–4766.
19. Huo, H.-H.; Zhang, H.-K.; Xia, X.-E.; Huang, P.-Q. A Formal Enantioselective Total Synthesis of FR901483. *Org. Lett.* **2012,** *14,* 4834–4837.
20. Huo, H.-H.; Xia, X.-E.; Zhang, H.-K.; Huang, P.-Q. Enantioselective Total Syntheses of (−)-FR901483 and (+)-8-*Epi*-Fr901483. *J. Org. Chem.* **2013,** *78,* 455–465.
21. Gaunt, M. J.; Reich, D.; Trowbridge, A. Rapid Syntheses of (−)-FR901483 and (+)-TAN1251C Enabled by Complexity-generating Photocatalytic Olefin Hydroaminoalkylation. *Angew. Chem.* **2019,** *132* (6), 2276–2281.

CHAPTER

# 12

# Synthetic approach to the TAN1251 alkaloids

## 12.1 Kawahara's total synthesis of TAN1251A (2002)

See (Scheme 167).

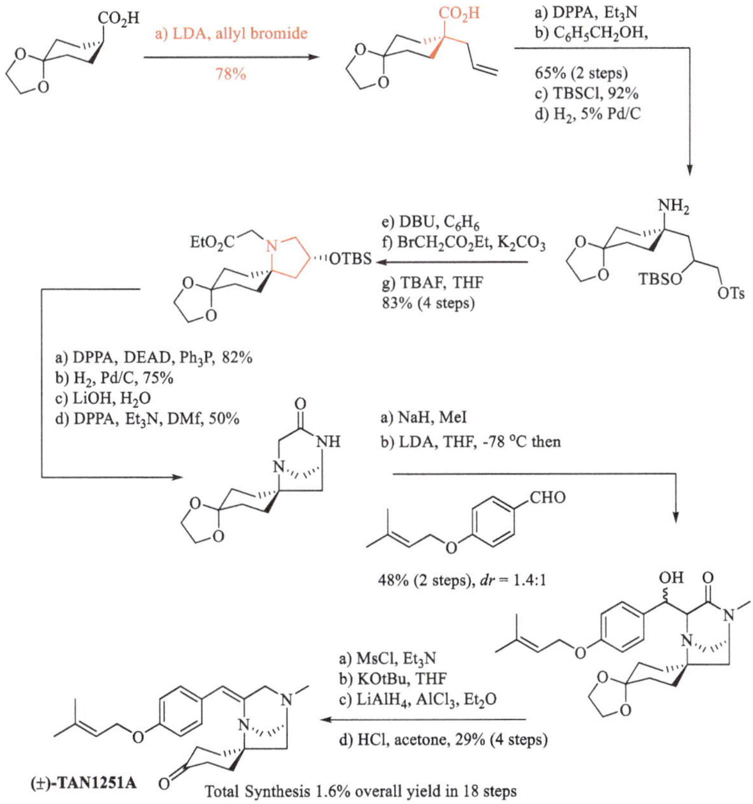


SCHEME 167 Total synthesis of TAN1251A.[1]


https://doi.org/10.1016/B978-0-12-822262-1.00012-3

## 12.2 Kawahara's total synthesis of (−)-TAN1251A (2002)

See (Scheme 168).

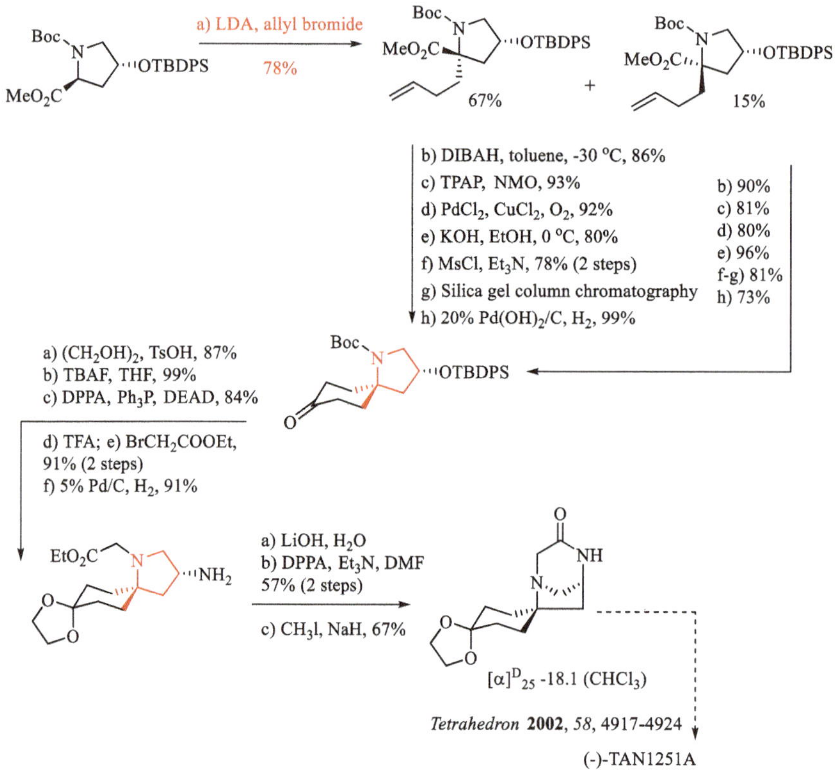


SCHEME 168 Total synthesis of (−)-TAN1251A.[2]

## 12.3 Wardrop's formal synthesis of (−)-TAN1251A (2001)

See (Scheme 169).

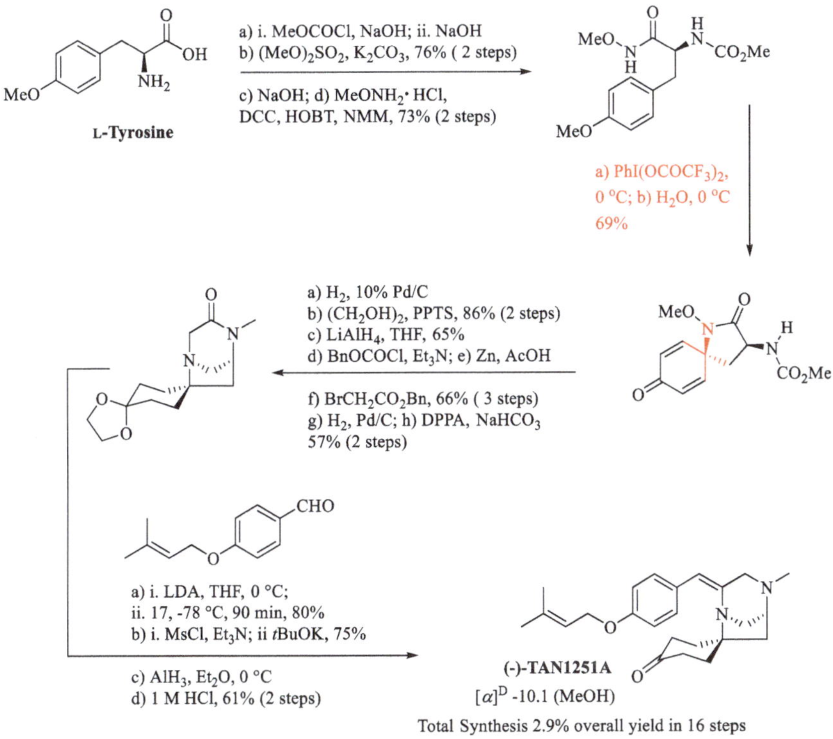


**SCHEME 169** Formal synthesis of (−)-TAN1251A via nitrenium ion–induced spirocyclization.[3]

## 12.4 Snider's biomimetic total syntheses of TAN1251A–D (2000)

See (Scheme 170).

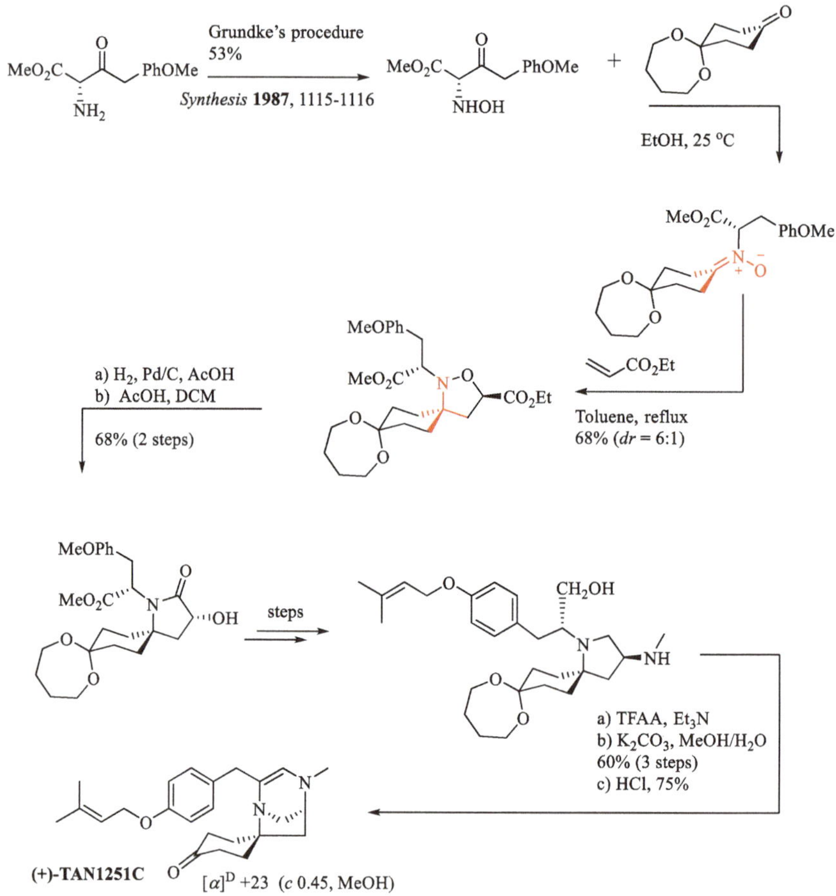


SCHEME 170 Biomimetic total syntheses of (-)–TAN1251A, (+)–TAN1251B, (+)–TAN1251C, and (+)–TAN1251D.

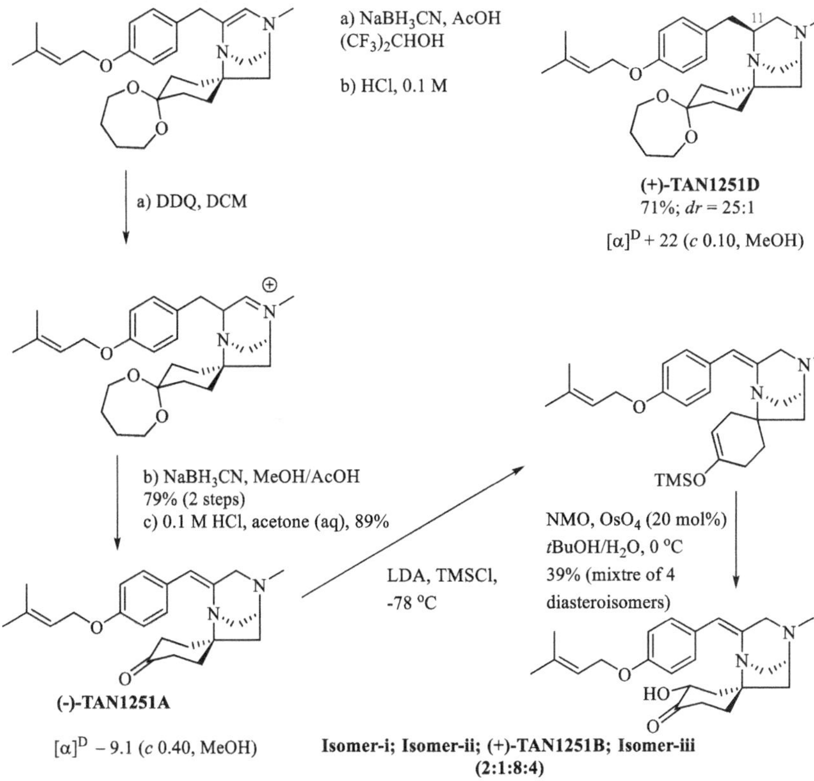


SCHEME 170 Cont'd.

## 12.5 Ciufolini's approach via oxidative cyclization (2001)

See (Scheme 171).

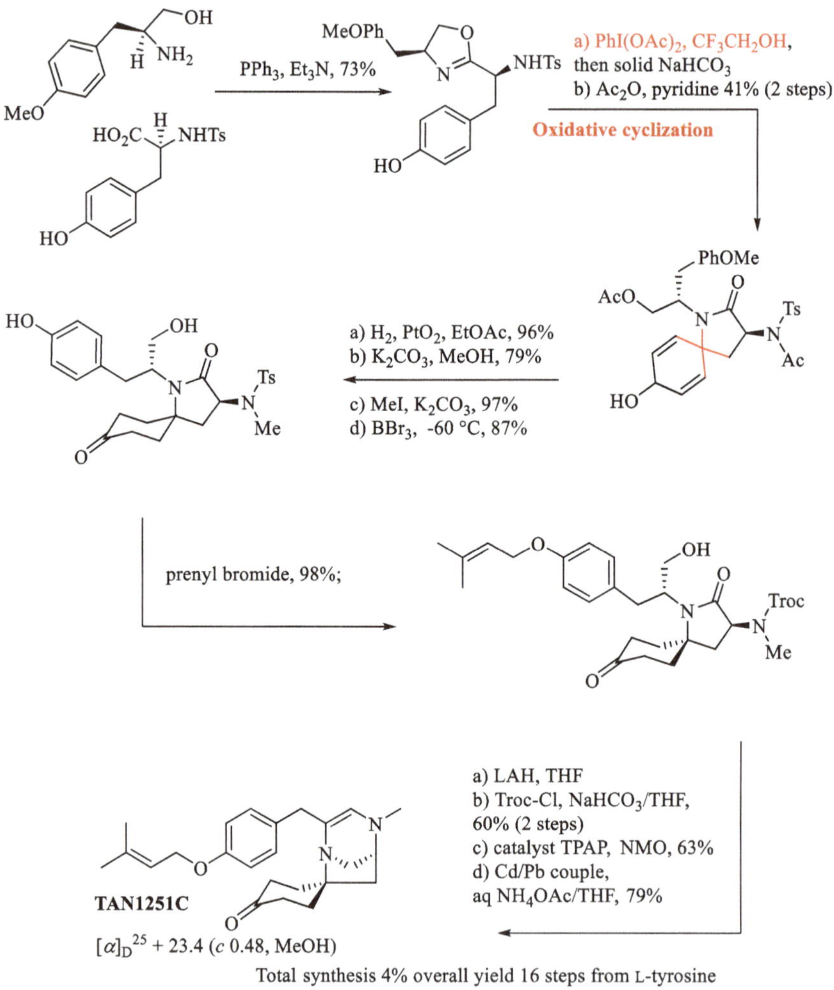


SCHEME 171 Synthetic studies via oxidative cyclization.[5]

## 12.6 Honda's formal synthesis (−)-TAN1251A (2002)

See (Scheme 172).

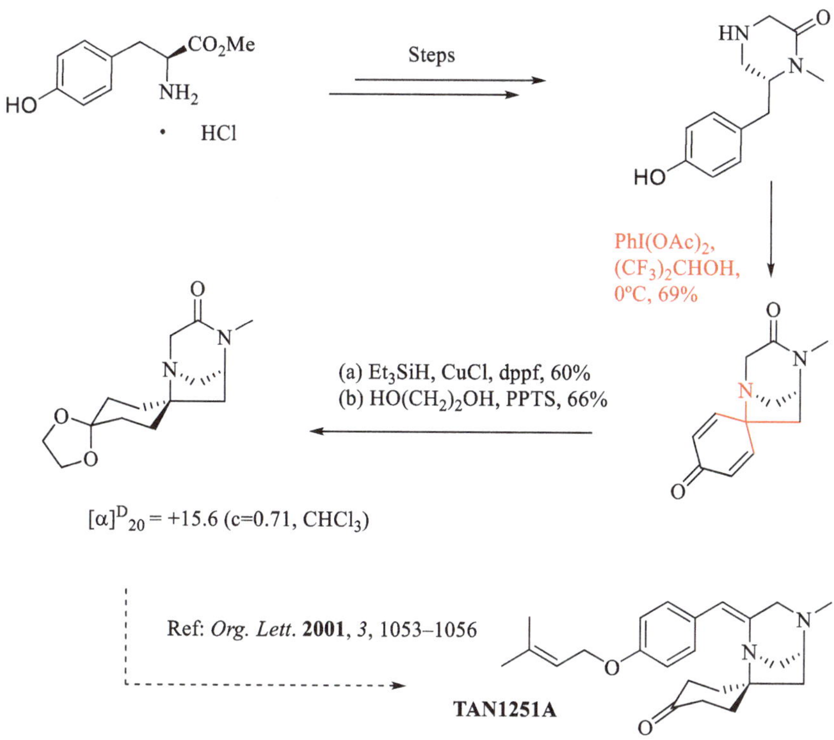


SCHEME 172 Formal synthesis of enantiopure (−)-TAN1251A via aromatic oxidation.[6]

## 12.7 Honda's enantiospecific total synthesis of TAN1251C and D (2002)

See (Scheme 173).

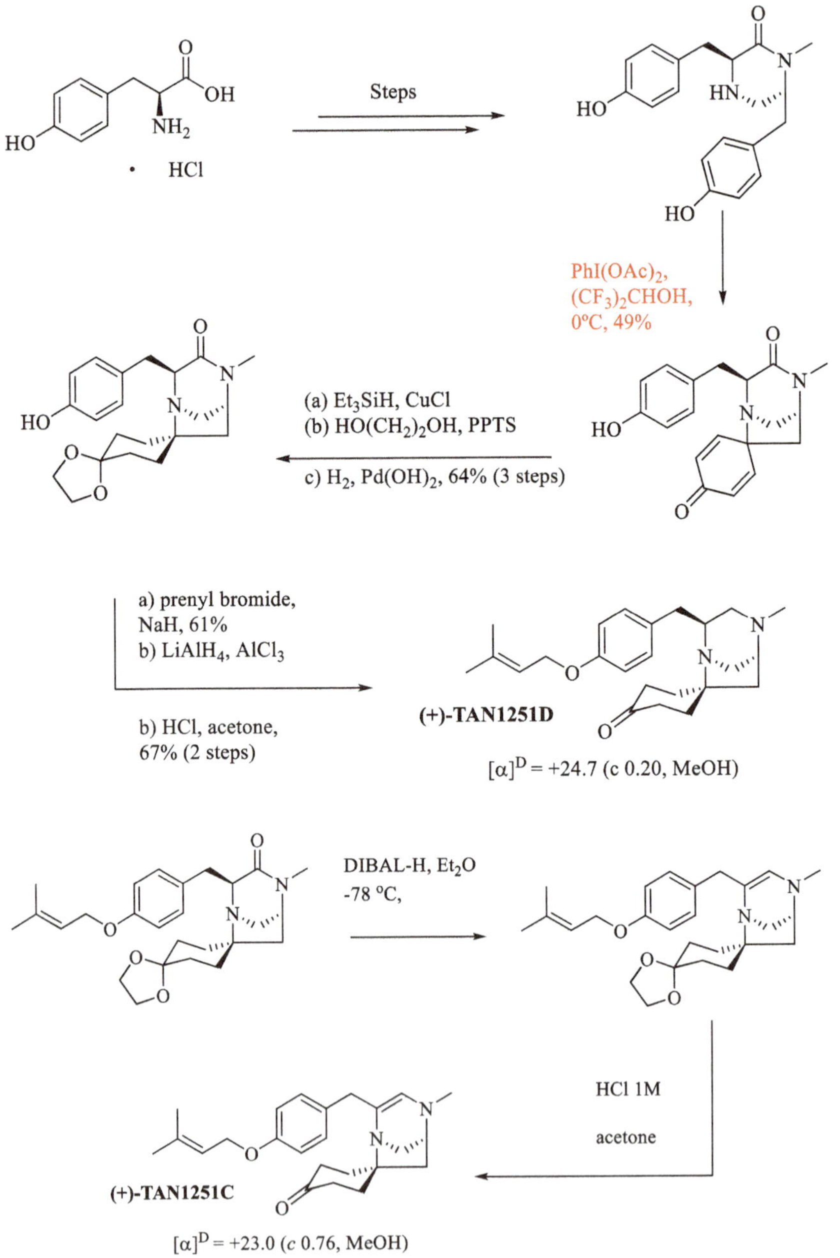


**SCHEME 173** Enantiospecific total synthesis of TAN1251C and TAN1251D via aromatic oxidation reaction of secondary amine.[7]

## 12.8 Hayes's enantioselective total synthesis of (−)-TAN1251A (2000)

See (Scheme 174).

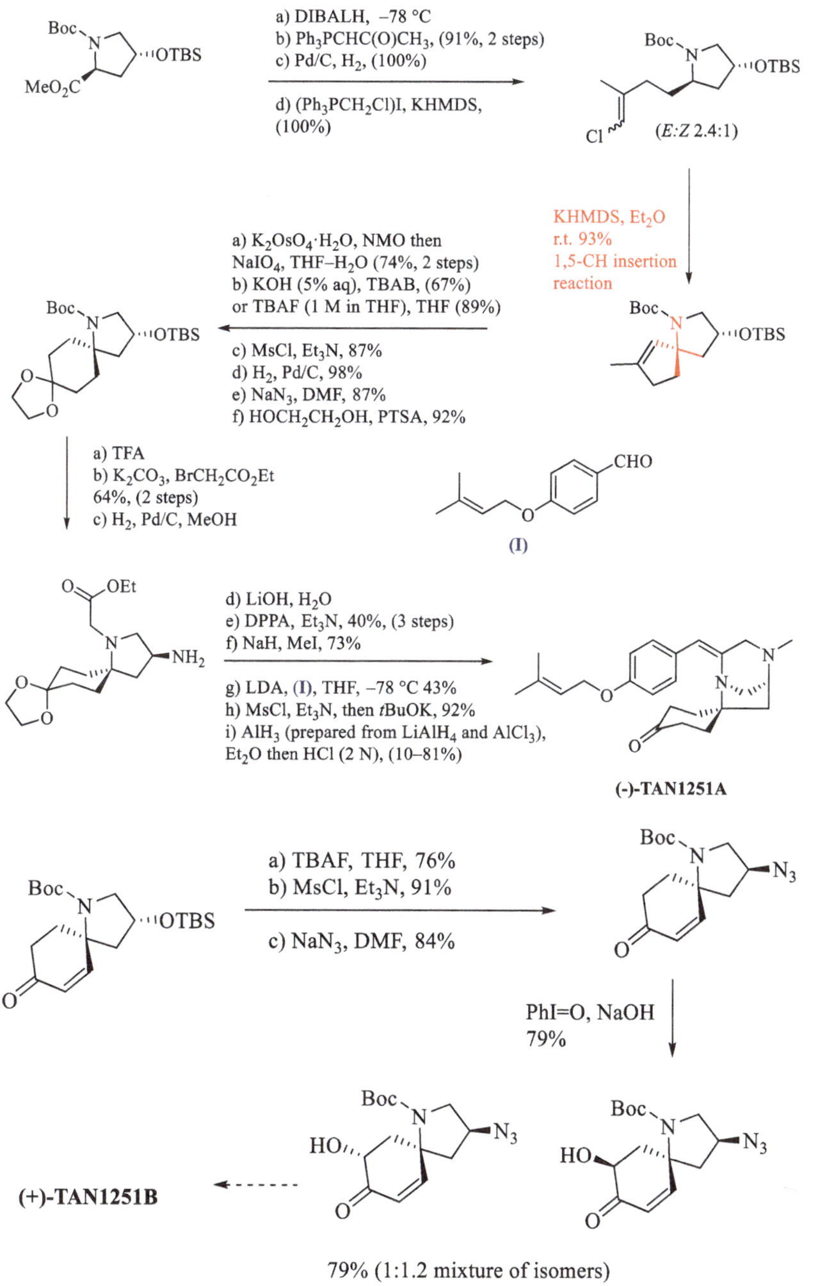


**SCHEME 174** Enantioselective total synthesis of (−)-TAN1251A via alkylidene 1, 5-CH insertion reaction.[8]

## 12.9 Peiqiang's enantioselective synthesis key core of TAN1251C

See (Scheme 175).

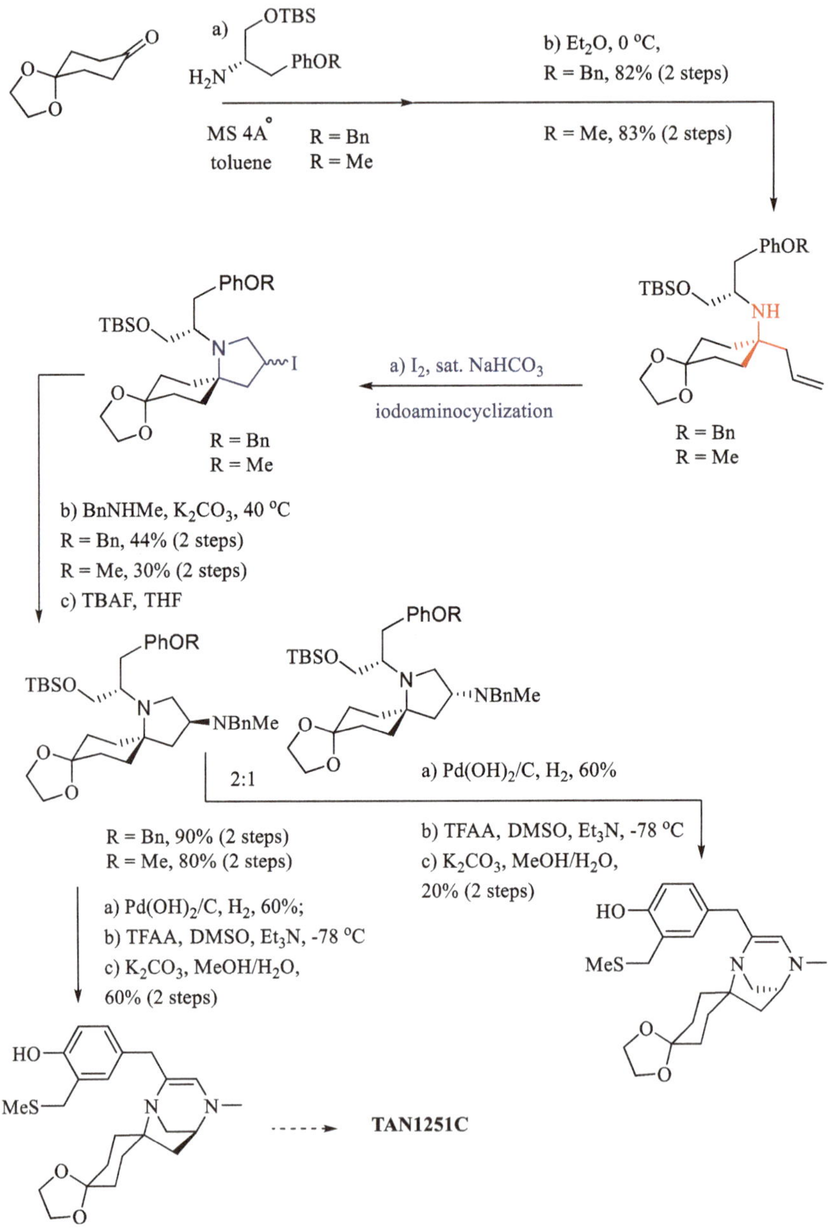


SCHEME 175 Enantioselective synthesis Of diazatricyclic core of alkaloid TAN1251C via an iodoaminocyclization reaction.[9]

## 12.10 Kan's total synthesis of TAN1251C (2017)

See (Scheme 176).

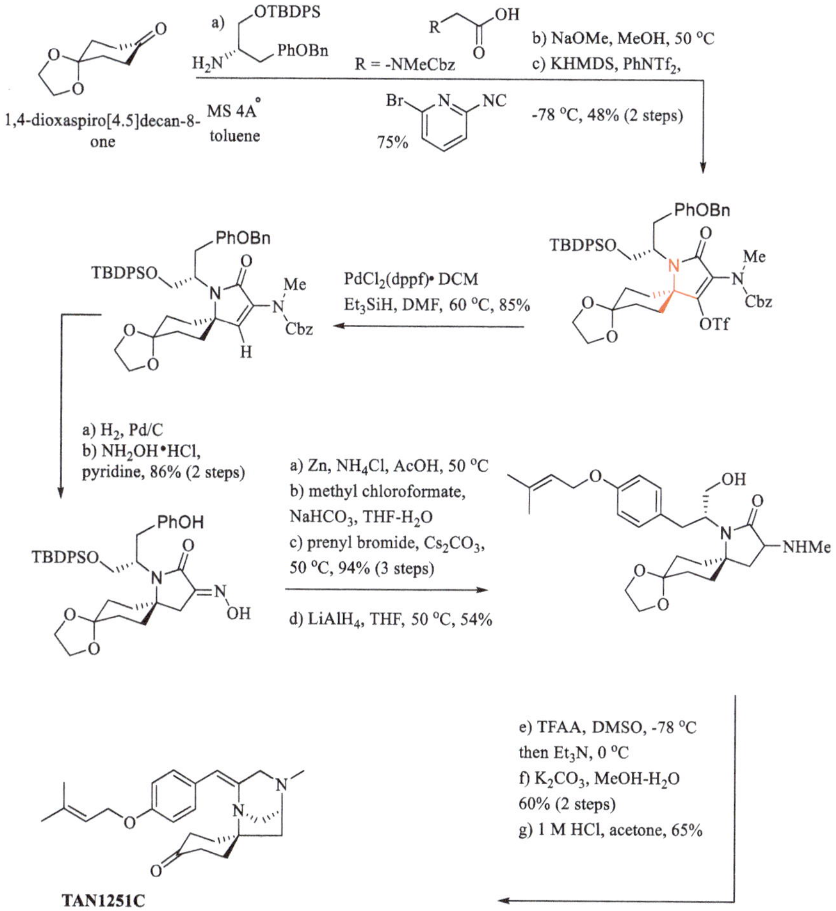


$[\alpha]^{D}_{25}$ + 23.8 (*c* 0.37, MeOH)

Total synthesis 5.2% overall yield in 13 steps from 1,4-dioxaspiro[4.5]decan-8-one

**SCHEME 176** Total synthesis of TAN1251C via diastereoselective construction of the azaspiro skeleton.[10]

## References

1. Nagumo, S.; Nishida, A.; Yamazaki, C.; Matoba, A.; Murashige, K.; Kawahara, N. Total Synthesis of Antimuscarinic Alkaloid,(±)-TAN1251A. *Tetrahedron* **2002,** *58,* 4917–4924. Nagumo, S.; Nishida, A.; Yamazaki, C.; Murashige, K.; Kawahara, N., Total synthesis of (±)-TAN1251A. *Tetrahedron Lett.* **39, 1998,** 4493–4496.
2. Nagumo, S.; Matoba, A.; Ishii, Y.; Yamaguchi, S.; Akutsu, N.; Nishijima, H.; Nishida, A.; Kawahara, N. Synthesis of (−)-TAN1251A Using 4-Hydroxy-L-Proline as a Chiral Source. *Tetrahedron* **2002,** *58,* 9871–9877.
3. Wardrop, D. J.; Basak, A. N-Methoxy-N-acylnitrenium Ions: Application to the Formal Synthesis of (−)-TAN1251A. *Org. Lett.* **2001,** *3,* 1053–1056.
4. Snider, B. B.; Lin, H. Biomimetic Total Syntheses of (−)-TAN1251A,(+)-TAN1251B,(+)-TAN1251C, and (+)-TAN1251D. *Org. Lett.* **2000,** *2* (5), 643–646.
5. Ousmer, M.; Braun, N. A.; Bavoux, C.; Perrin, M.; Ciufolini, M. A. Total Synthesis of Tricyclic Azaspirane Derivatives of Tyrosine: FR901483 and TAN1251C. *J. Am. Chem. Soc.* **2001,** *123,* 7534–7538.
6. Mizutani, H.; Takayama, J.; Soeda, Y.; Honda, T. Facile Synthesis of Enantiopure (−)-TAN1251A. *Tetrahedron Lett.* **2002,** *43,* 2411–2414.
7. Mizutani, H.; Takayama, J.; Honda, T. Enantiospecific Total Synthesis of TAN1251C and TAN1251D. *Synlett* **2005,** *2005* (02), 328–330.
8. Auty, J. M.; Churcher, I.; Hayes, C. J. An Enantioselective Formal Total Synthesis of (-)-TAN1251A. *Synlett* **2004,** *2004* (08), 1443–1445. Asari, A.; Hayes, C. J., A Study Towards the Total Synthesis of TAN1251B. Sains Malaysiana **38**, 2009, 869–872.
9. Zhang, H.; Lin, Z.; Huang, H.; Huo, H.; Huang, Y.; Ye, J.; Huang, P. Enantioselective Synthesis of the Diazatricyclic Core of Alkaloid TAN1251C *via* an Iodoaminocyclization Reaction. *Chin. J. Chem.* **2010,** *28,* 1717–1724.
10. Nagasaka, Y.; Shintaku, S.; Matsumura, K.; Masuda, A.; Asakawa, T.; Inai, M.; Egi, M.; Hamashima, Y.; Ishikawa, Y.; Kan, T. Total Synthesis of TAN1251C *via* Diastereoselective Construction of the Azaspiro Skeleton. *Org. Lett.* **2017,** *19,* 3839–3842.

CHAPTER

# 13

# (1*S*,3*R*)-1-Aminocyclopentane-1,3-diarboxylic acid (ACPD)

## 13.1 Abstract

(1*S*,3*R*)-1-Aminocyclopentane-1,3-diarboxylic acid (ACPD) is a conformationally constrained analog of L-glutamate that is used as a chemical messenger in most of excitatory synapses in the central nervous system.[1] Most of the excitatory synapses in the central nervous system use glutamate (Glu) as a chemical messenger (Fig. 12).

ACPD has therapeutic potential for neurodegenerative disorders and serves as a selective antagonist of metabotropic glutamate receptors (mGluRs), which are coupled with G proteins to mediate various transduction mechanisms.[2] Generally, the glutamate receptors have been classified into two main groups termed as iGluRs (ionotropic glutamate receptors) and mGluRs. ACPD serves as a potential candidate for regulation of glutamate receptor function to treat CNS disorders. ACPD has cyclopentane core bearing quaternary center next to nitrogen atom. Interesting biological potential and different methods to build *N*-bearing quaternary center to complete stereoselective synthesis of ACPD are included in this section.

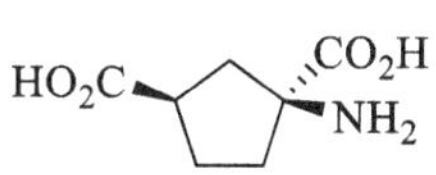


(1*S*,3*R*)-1-Aminocyclopentane-1,3-diarboxylic Acid
(ACPD)

**FIGURE 12** Structure of ACPD. ACPD, (1*S*,3*R*)-1-aminocyclopentane-1,3-diarboxylic acid.


https://doi.org/10.1016/B978-0-12-822262-1.00013-5

## 13.2 Ma's total synthesis of (1*S*,3*R*)-1-aminocyclopentane-1,3-diarboxylic acid (1997)

See (Scheme 177).

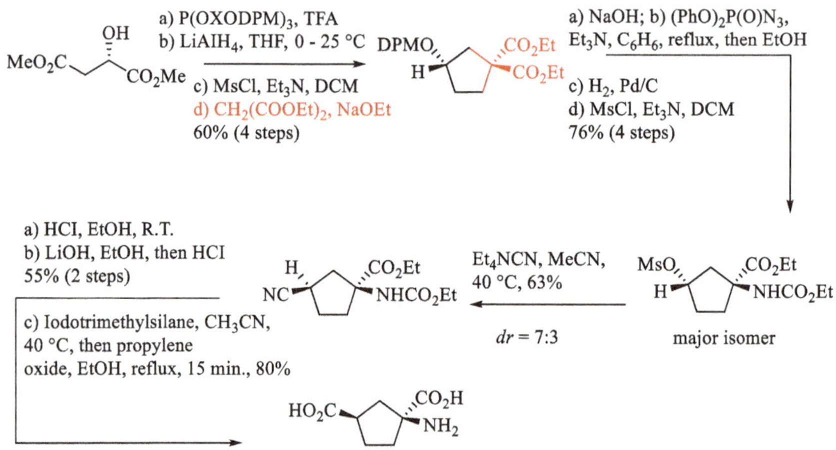


**ACPD**

**SCHEME 177** Total synthesis of ACPD.[3] ACPD, (1*S*,3*R*)-1-aminocyclopentane-1,3-diarboxylic acid.

## 13.3 Hodgson's total (1*S*,3*R*)-1-aminocyclopentane-1,3-diarboxylic acid synthesis via hydroboration

See (Scheme 178).

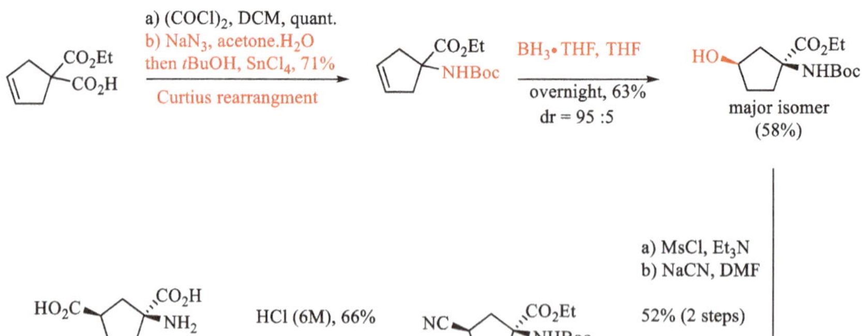


**ACPD**

**SCHEME 178** Total ACPD synthesis via hydroboration.[4] ACPD, (1*S*,3*R*)-1-aminocyclopentane-1,3-diarboxylic acid.

## 13.4 Hayes's total (1*S*,3*R*)-1-aminocyclopentane-1,3-diarboxylic acid synthesis via 1,5-CH insertion

See (Scheme 179).

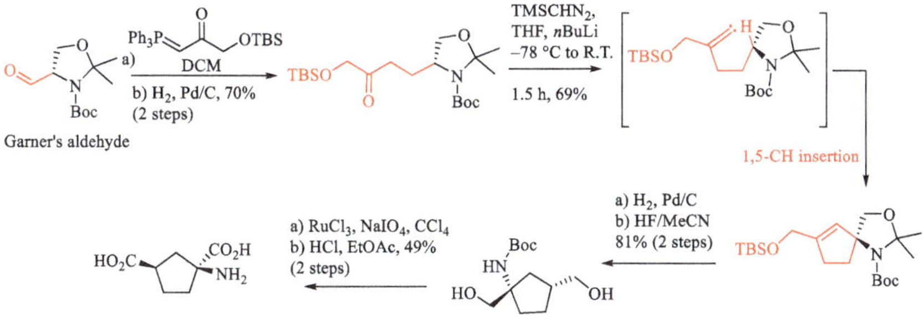


**SCHEME 179** Total ACPD synthesis via 1,5-CH insertion.[5] ACPD, (1*S*,3*R*)-1-aminocyclopentane-1,3-diarboxylic acid.

# References


1. Pin, J.-P.; Duvoisin, R. The Metabotropic Glutamate Receptors: Structure and Functions. *Neuropharmacology* **1995,** *34,* 1–26.
2. Knöpfel, T.; Gasparini, F. Metabotropic Glutamate Receptors: Potential Drug Targets. *Drug Discov. Today* **1996,** *1,* 103–108.
3. Ma, D.; Ma, J.; Dai, L. Stereospecific Synthesis of (1*S*, 3*R*)-1-Aminocyclopentane-1, 3-dicarboxylic Acid, a Selective Agonist of Metabotropic Glutamate Receptors. *Tetrahedron: Asymmetry* **1997,** *8* (6), 825–827.
4. Hodgson, D. M.; Thompson, A. J.; Wadman, S.; Keats, C. J. On the Possibility of Carbamate-Directed Hydroboration. An Approach to the Asymmetric Synthesis of 1-aminocyclopentane-1, 3-dicarboxylic Acid. *Tetrahedron* **1999,** *55* (35), 10815–10834.
5. Bradley, D. M.; Mapitse, R.; Thomson, N. M.; Hayes, C. J. Enantioselective Synthesis of the Excitatory Amino Acid (1 S, 3 R)-1-Aminocyclopentane-1, 3-dicarboxylic Acid. *J. Org. Chem.* **2002,** *67* (22), 7613–7617.

CHAPTER

# 14

# Tetrodotoxin

## 14.1 Abstract

Tetrodotoxin (TTX) was isolated from puffer fish, used as speciality in Japanese cuisine, in 1909 from the ovaries and livers of fish, while its structure was elucidated in 1964 on the basis of Tsuda, Woodward, and Goto work that reported independently to identify and characterize the compound with polyfunctionalized dioxaadamantane skeleton, a cyclic guanidine containing a hemiaminal moiety with nine contiguous stereogenic centers.[1] The TTX was found to be a potent selective blocker of voltage-gated sodium ($Na^{+}$) channels.[2] In 1970, Nitta et al. determine the absolute stereochemistry of TTS via X-ray crystallographic analysis of one of its derivatives.[3] In acidic conditions, TTX exits as inseparable mixture of orthoester and lactone; therefore, the purification of the tetrodotoxin in final steps is a challenging issue. Later, Yasumoto and Yotsu-Yamashita have isolated many natural analogs of TTX from puffer fish and newts.[4] Most of these natural derivatives consist of deoxy species and regard as biosynthetic intermediates or metabolites of tetrodotoxin. TTX with its potent biological activity and structure complexity have obtained much attention from chemist community to synthesize it (Fig. 13).

## 14.2 Du Bois's stereoselective synthesis of (−)-tetrodotoxin (2003)

See (Scheme 180).


https://doi.org/10.1016/B978-0-12-822262-1.00014-7

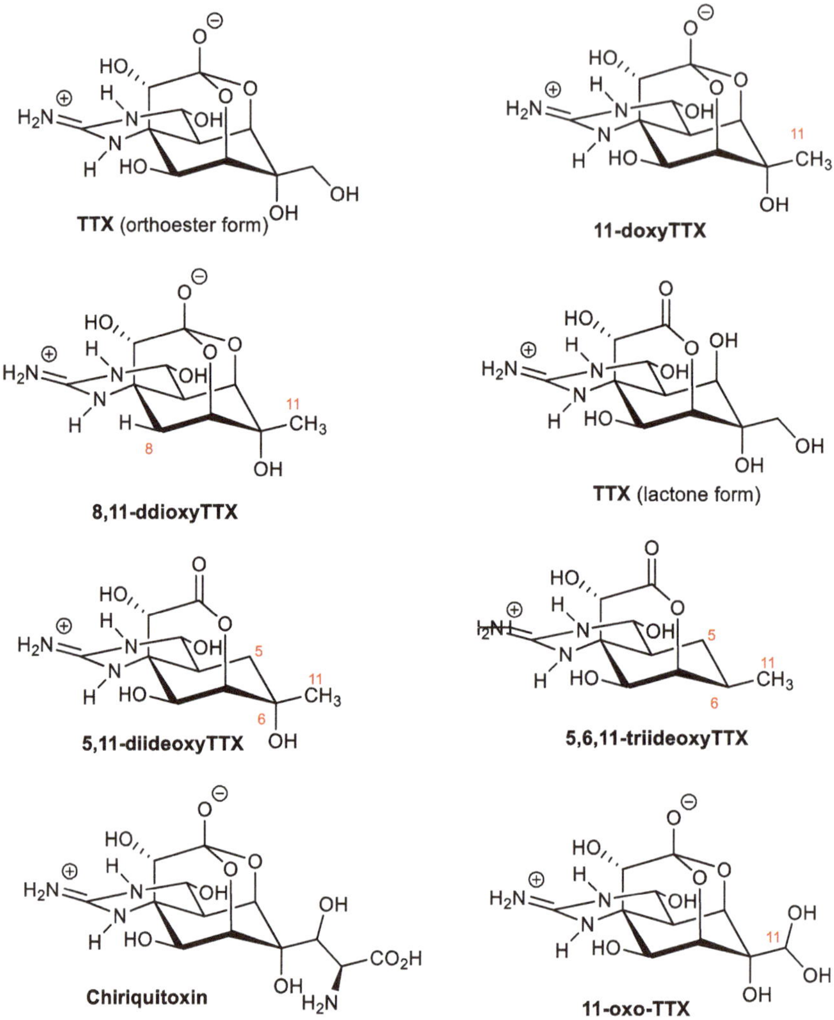


FIGURE 13 Structures of tetrodotoxin and its naturally occurring analogs.[5,6]

## 14.3 Isobe's first asymmetric total synthesis (2003)

See (Scheme 181).

## 14.4 Isobe's efficient total synthesis of tetrodotoxin (2004)

See (Scheme 182).

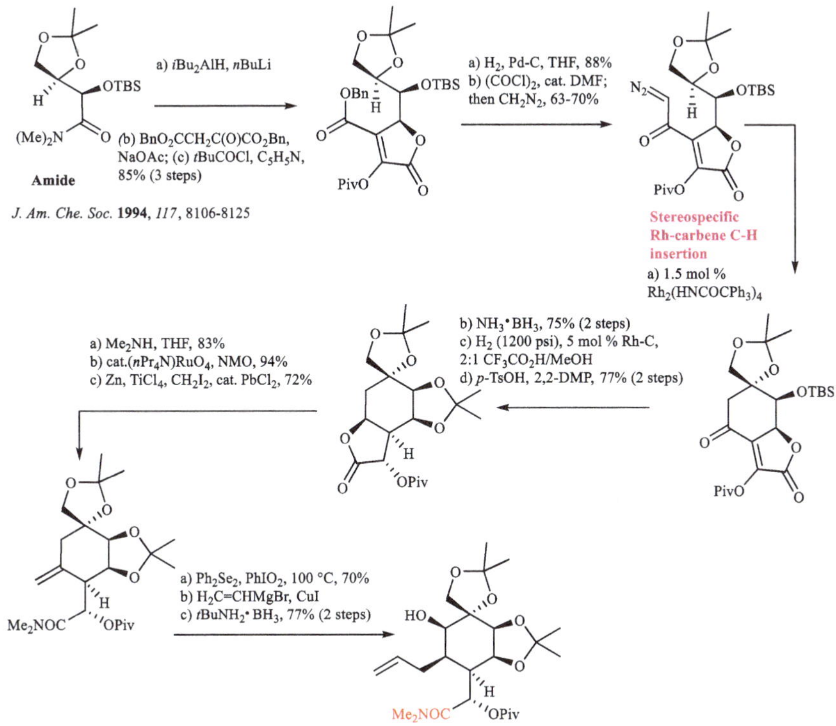


**SCHEME 180** Stereoselective synthesis of (−)-tetrodotoxin via Rh-catalyzed *C–H* insertion and stereospecific *C–H* amination.[7]

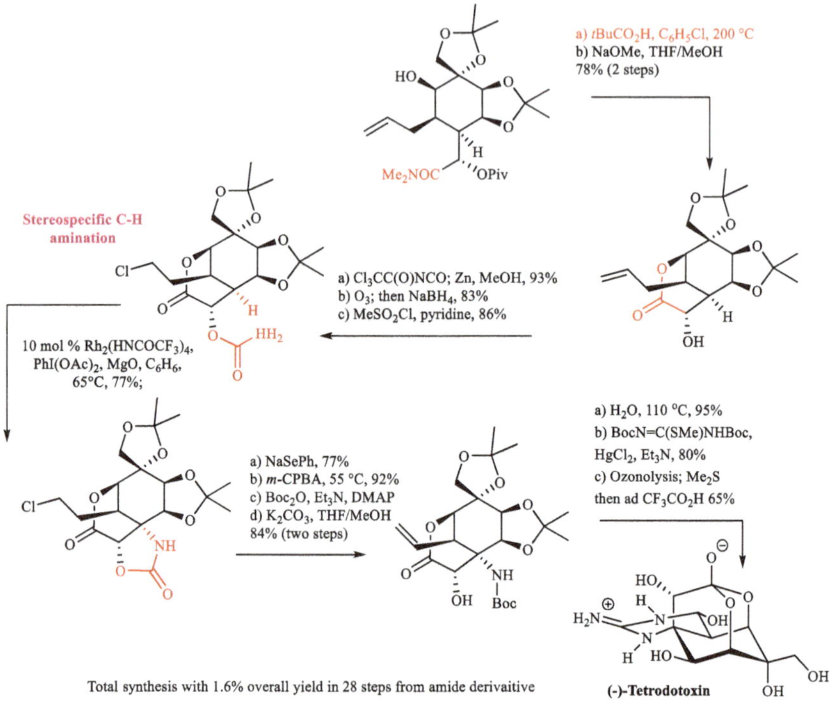


SCHEME 180 Cont'd.

## 14.5 Sato's stereocontrolled synthesis of (±)-tetrodotoxin (2005)

See (Scheme 183).

## 14.6 Ohfune's synthesis of (−)-5,6,11-trideoxytetrodotoxin and its 4-epimer (2006)

See (Scheme 184).

## 14.7 Sato's stereoselective synthesis of tetrodotoxin (2007)

See (Scheme 185).

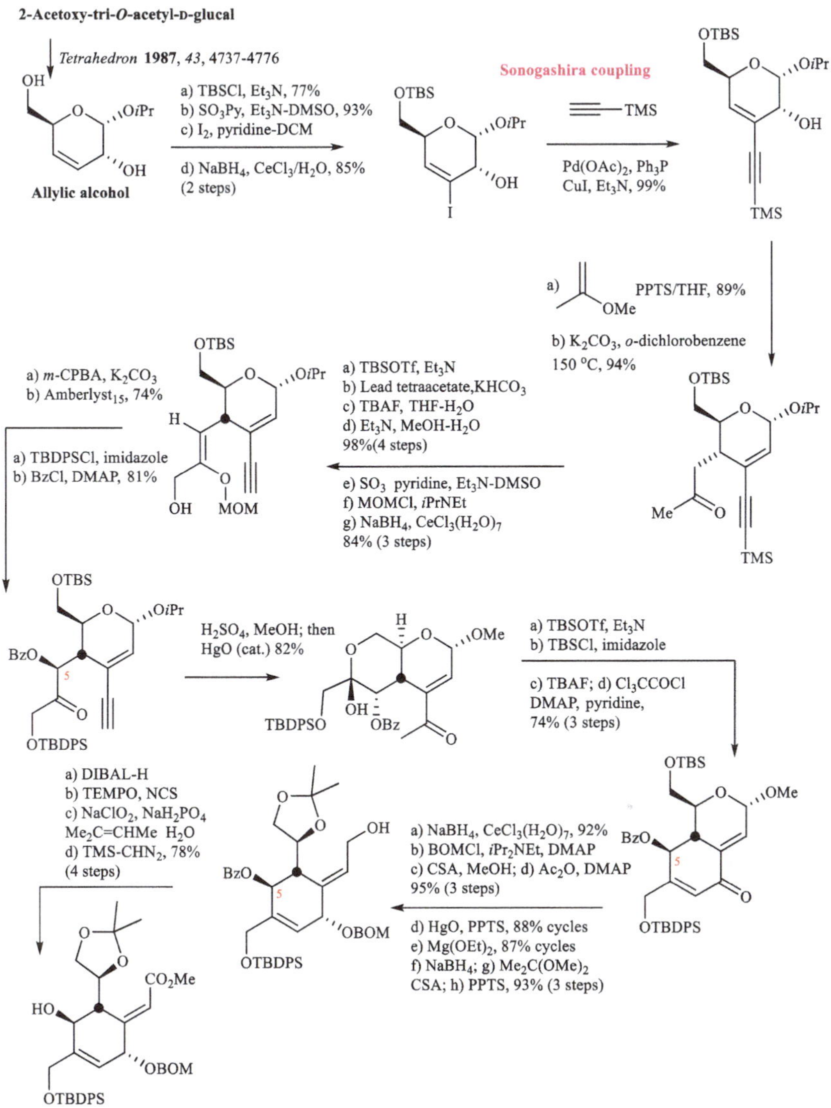


SCHEME 181 Asymmetric total synthesis via a new route from 2-acetoxy-tri-O-acetyl-D-glucal.[8]

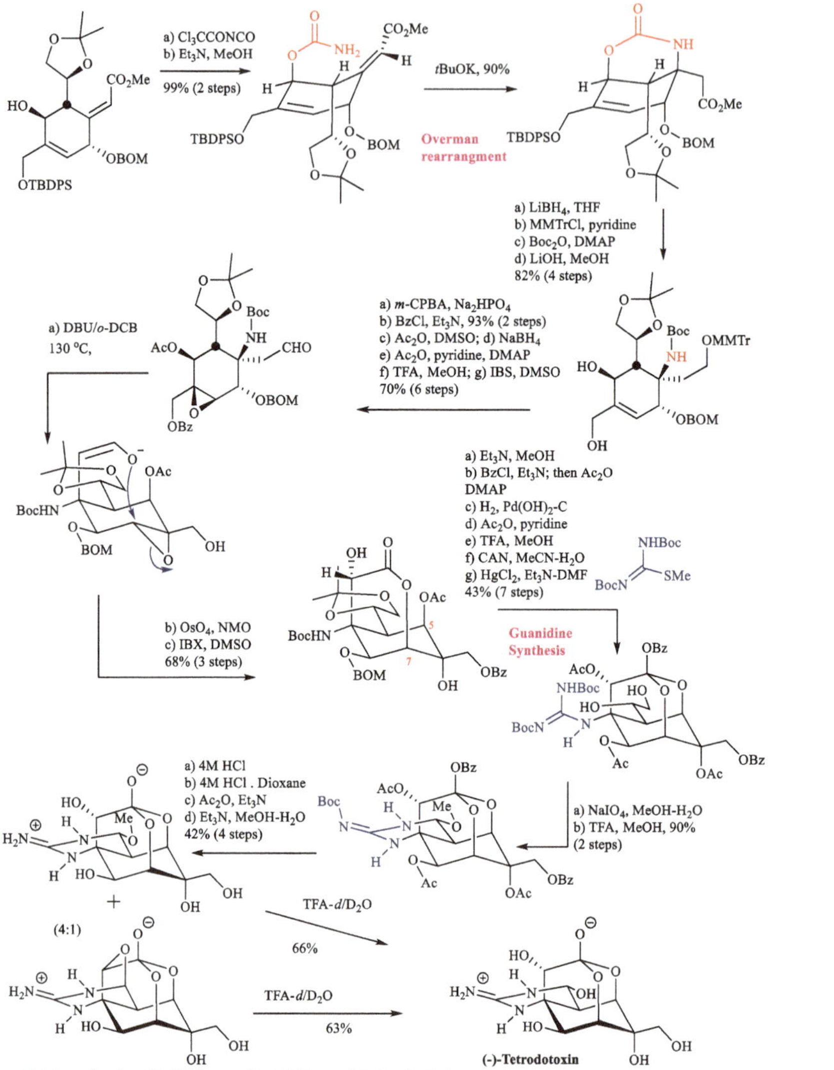


Total synthesis with 0.3% overall yield from Allylic alcohol

$[\alpha]_D^{28}$ +1 (*c* 0.08, 0.05 N AcOH-$H_2O$); Lit. -8.64 (c 8.55 in dilute AcOH) (*Pharm. Bull. Japan* **1953**, *1*, 112-113)

SCHEME 181 Cont'd.

## 14.8 Nishikawa's synthesis of (−)-5,11-dideoxytetrodotoxin (2013)[12]

See (Scheme 186).

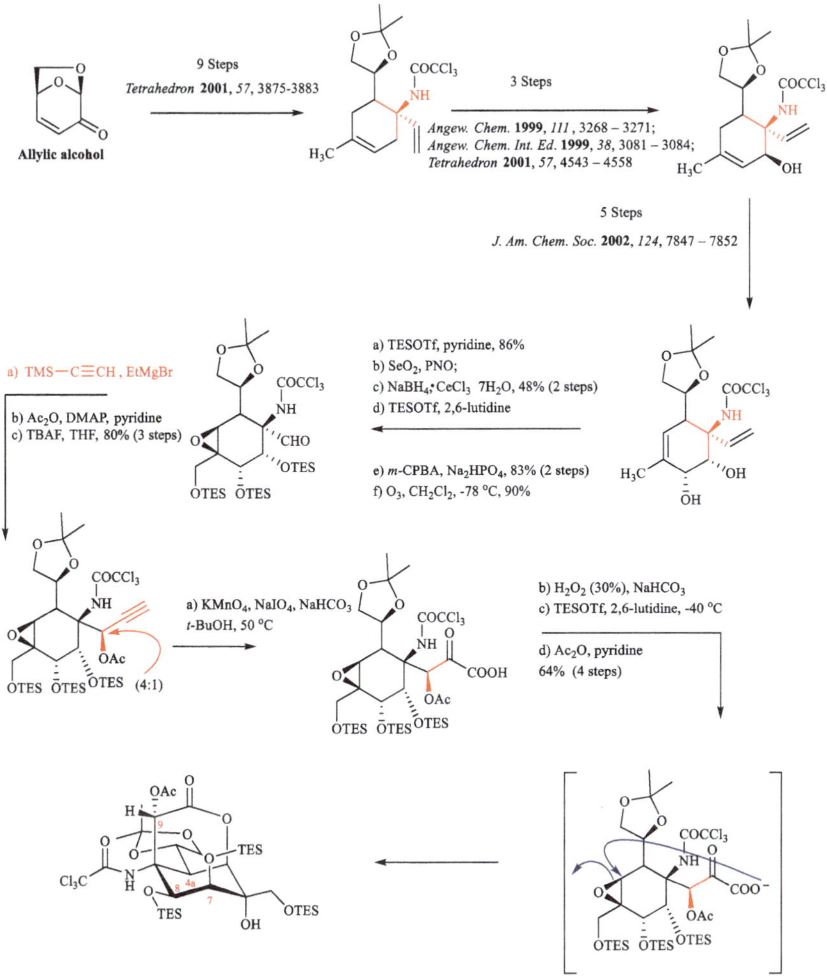


**SCHEME 182** Efficient total synthesis of optically active tetrodotoxin.[9]

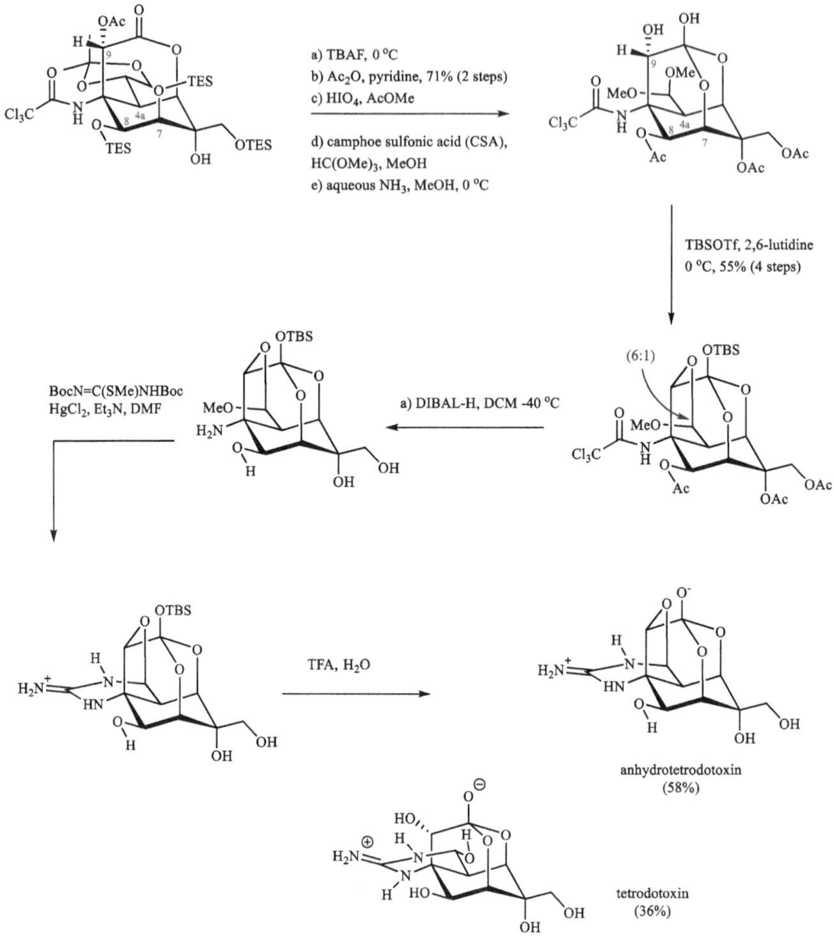


SCHEME 182 Cont'd.

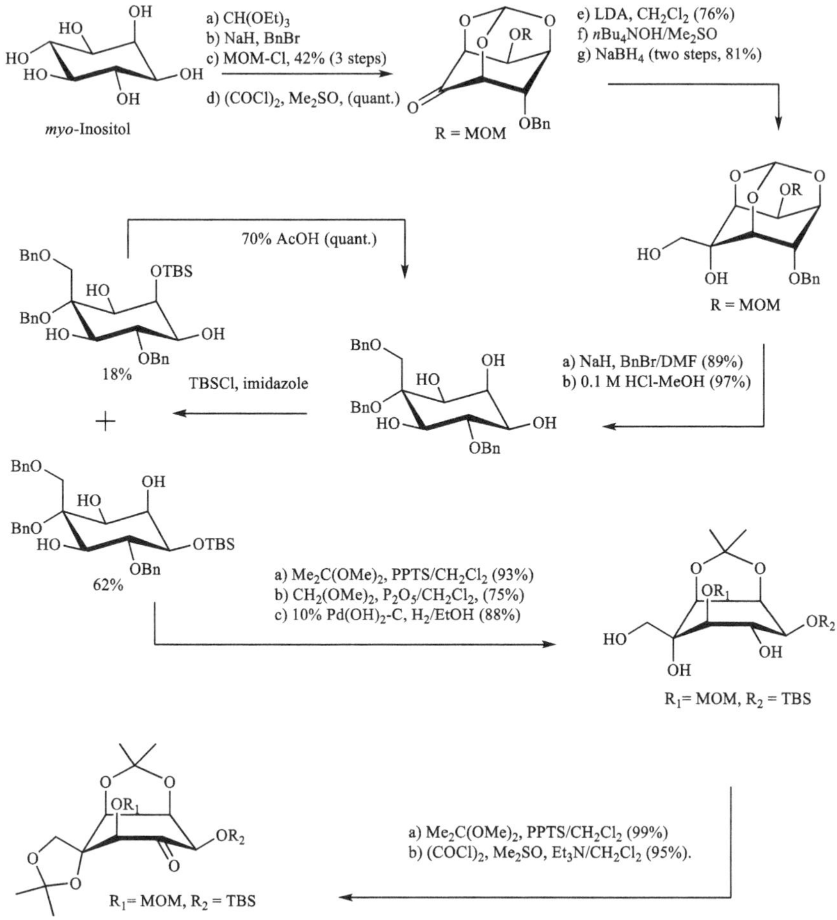


**SCHEME 183** Stereocontrolled synthesis of (±)-tetrodotoxin from *myo*-inositol.[10]

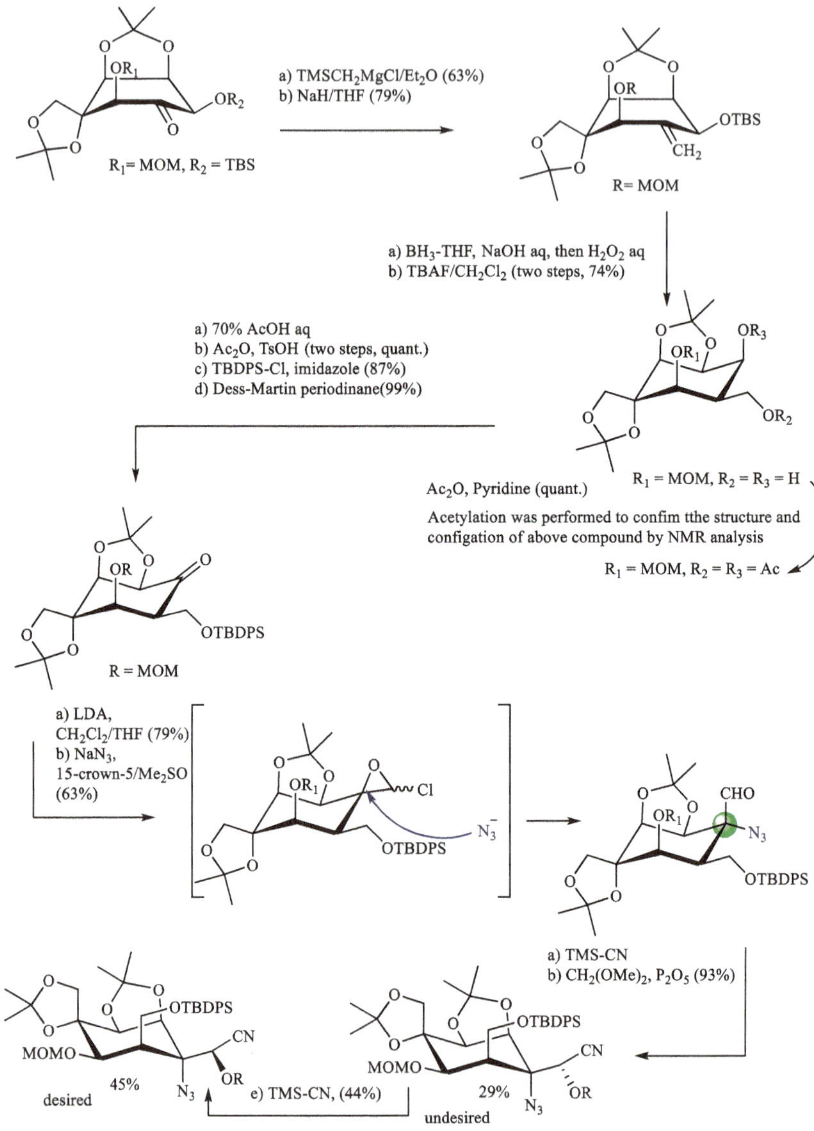


SCHEME 183 Cont'd.

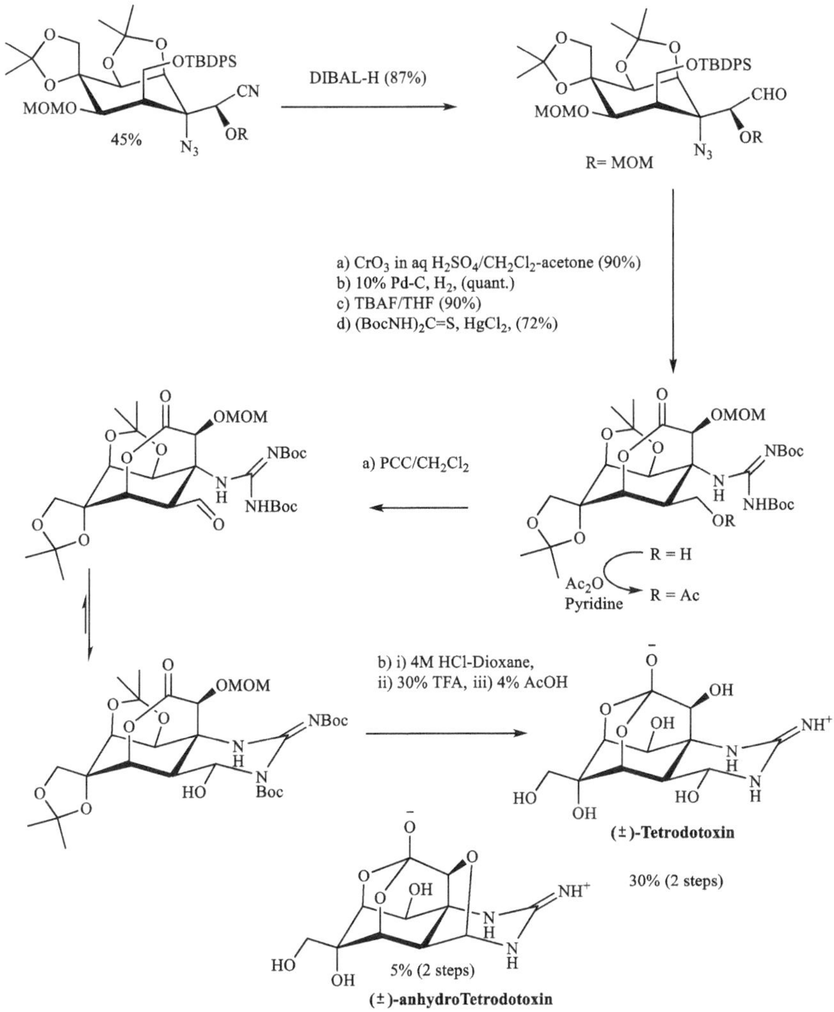


SCHEME 183 Cont'd.

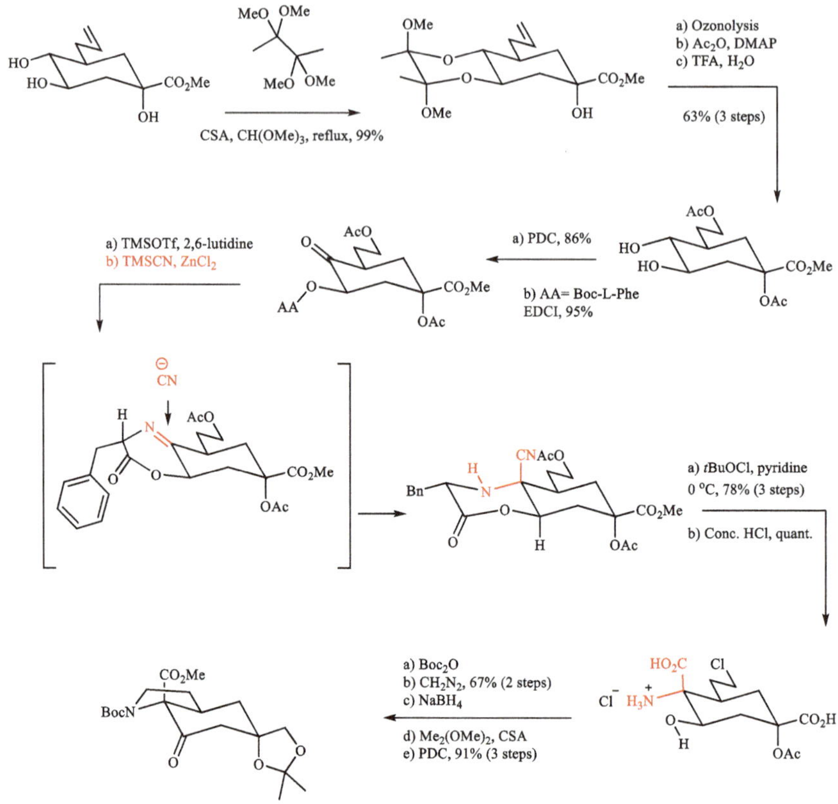


**SCHEME 184** Synthesis of (−)-5,6,11-tideoxytetrodotoxin and its 4-epimer.[11]

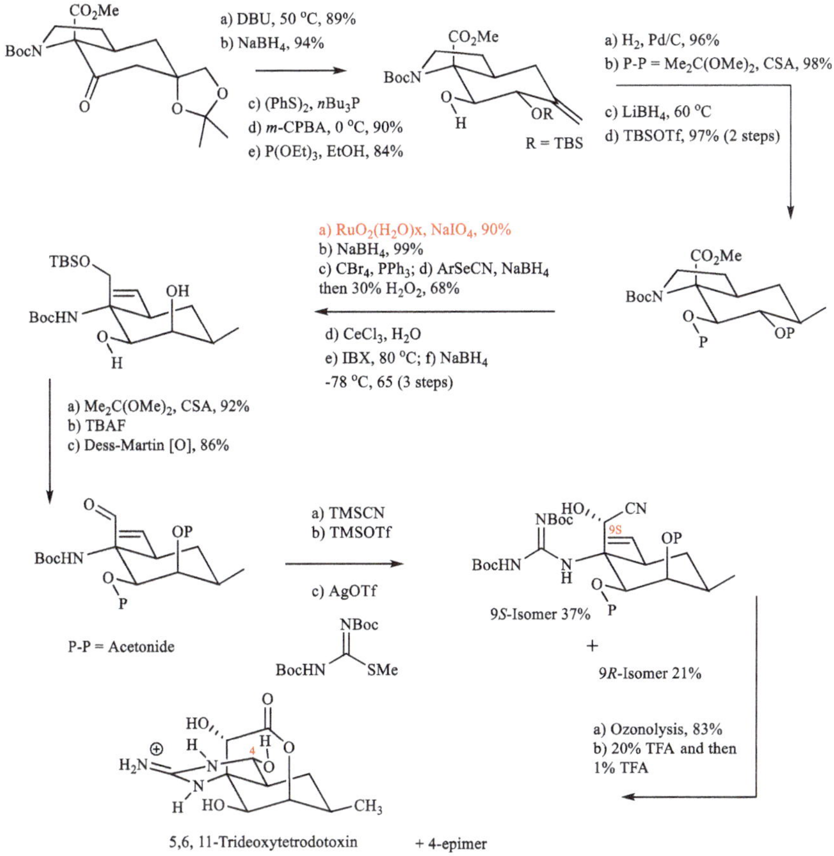


SCHEME 184 Cont'd.

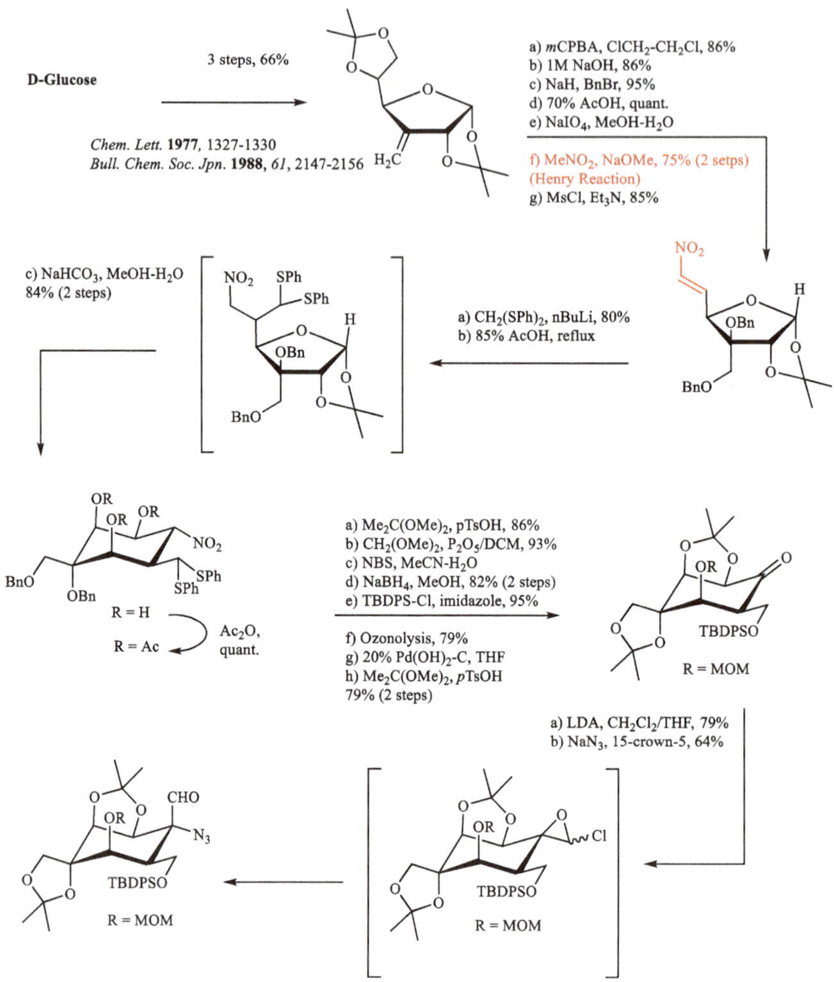


SCHEME 185 Tetrodotoxin synthesis from D-glucose.[9]

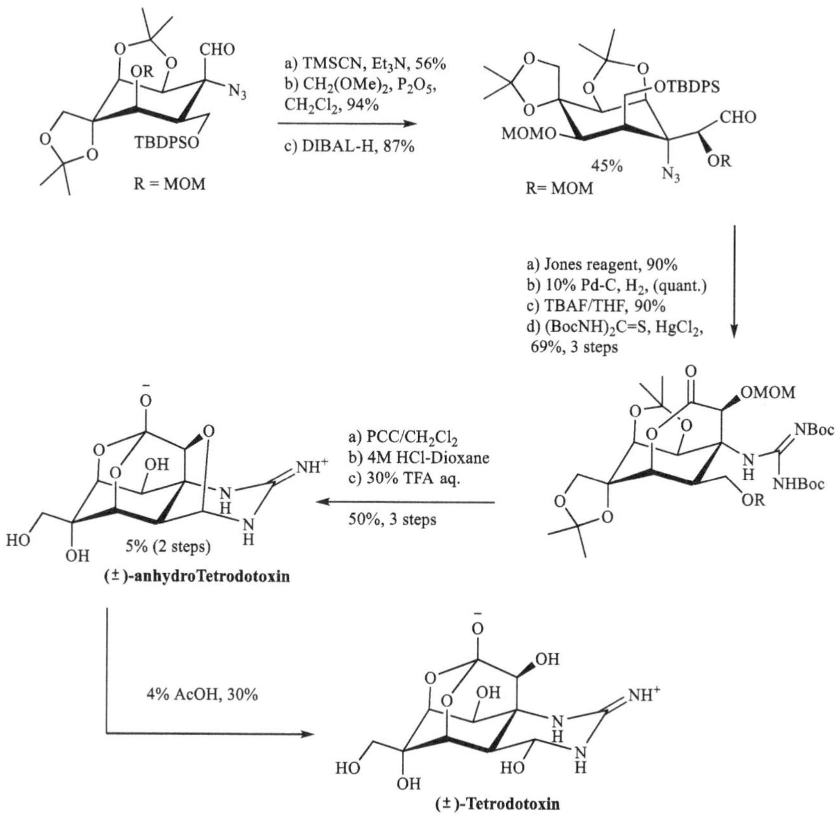


SCHEME 185 Cont'd.

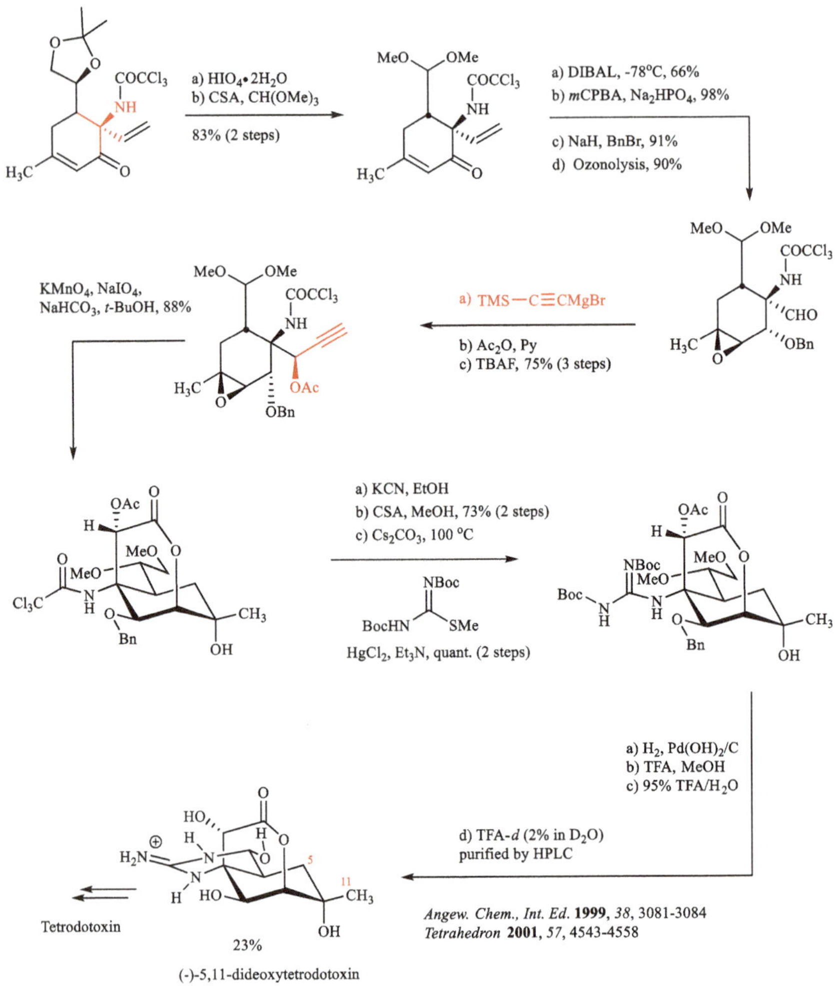


SCHEME 186 Synthesis of (−)-5,11-dideoxytetrodotoxin.

## 14.9 Ciufolin's formal synthesis of (±)-tetrodotoxin (2015)

See (Scheme 187).

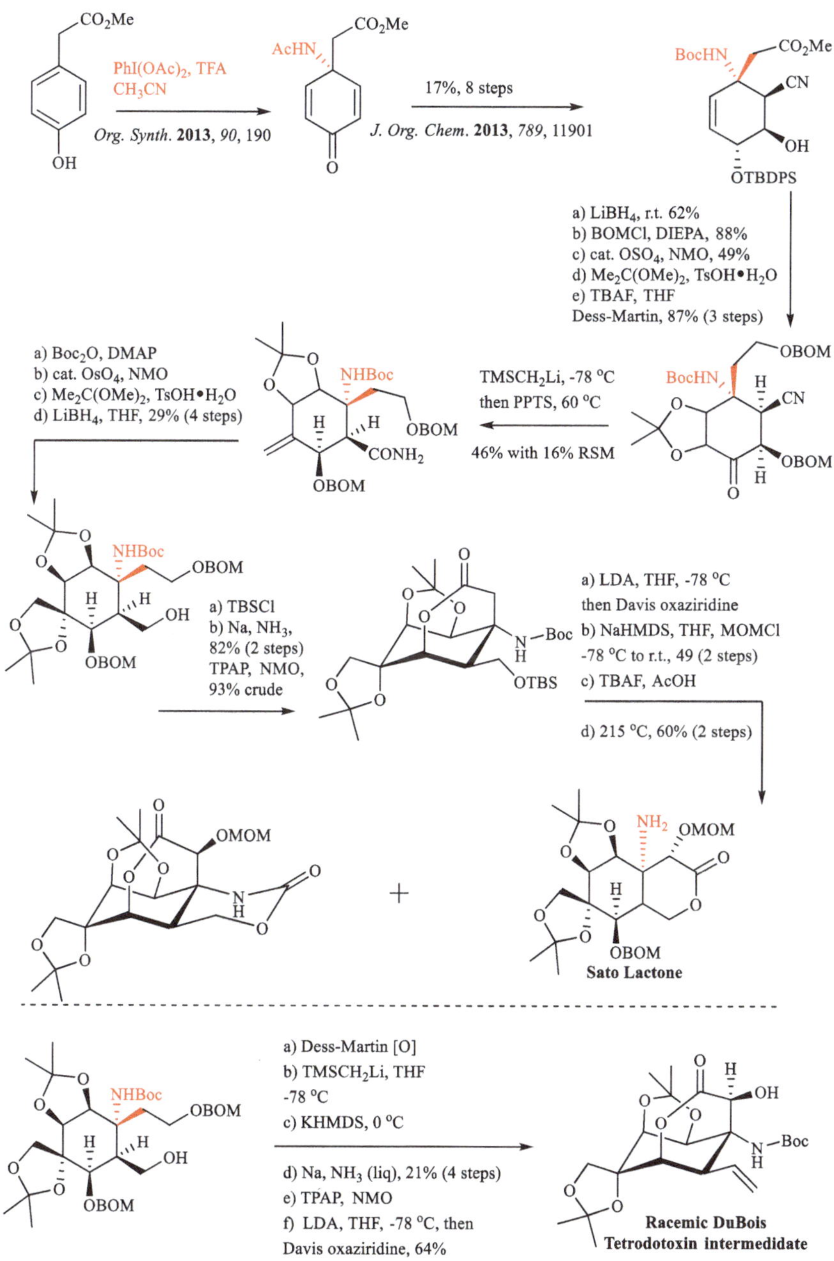


**SCHEME 187** Formal synthesis of (±)-tetrodotoxin via the oxidative amidation.[13]

## 14.10 Fukuyama's total synthesis of (−)-tetrodotoxin (2017)

See (Scheme 188).

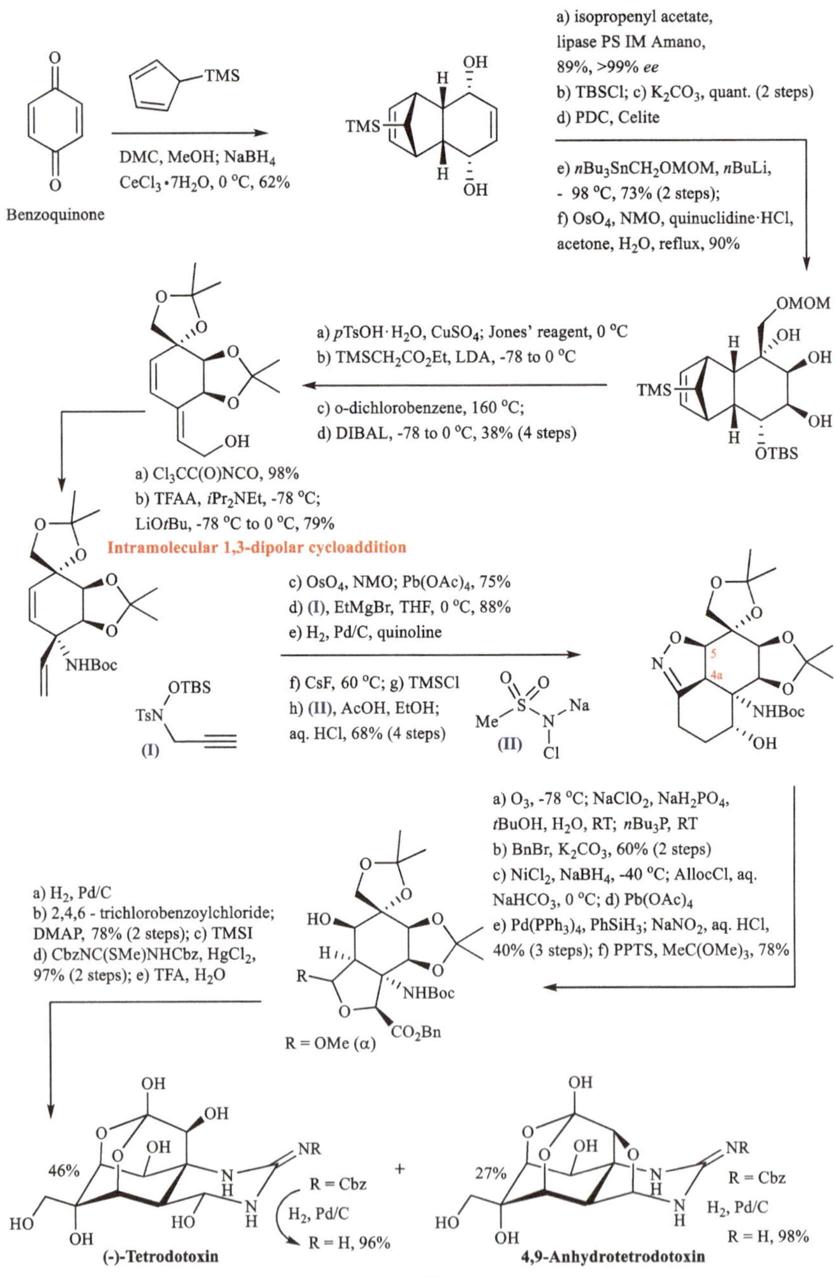


Overall Yield = 0.3% from Benzoquinone $[\alpha]^{23}_{D}$ -1.63° (*c* 0.25, 0.05 M AcOH)

SCHEME 188 Total synthesis of (−)-tetrodotoxin.[14]

# References


1. a) Goto, T.; Kishi, Y.; Takahashi, S.; Hirata, Y. *Tetrahedron* **1965,** *21,* 2059–2088.
   b) Tsuda, K.; Ikuma, S.; Kawamura, M.; Tachikawa, R.; Sakai, K.; Tamura, C.; Amakasu, O. *Chem. Pharm. Bull.* **1964,** *12,* 1357–1374.
   c) Woodward, R. B. *Pure Appl. Chem.* **1964,** *9,* 49–74.
2. a) Choudhary, G.; Yotsu-Yamashita, M.; Shang, L.; Yasumoto, T.; Dudley, S. C., Jr. *Biophys. J.* **2003,** *84,* 287–294.
   b) Narahashi, T. *J. Toxicol. Toxin ReV* **2001,** *20,* 67–84.
3. Furusaki, A.; Tomie, Y.; Nitta, I. *Bull. Chem. Soc. Jpn.* **1970,** *43,* 3325.
4. Yotsu-Yamashita, M. *J. Toxicol. Toxin Rev.* **2001,** *20,* 51.
5. Yotsu-Yamashita, M. Chemistry of Puffer Fish Toxin. *J. Toxicol. Toxin Rev.* **2001,** *20* (1), 51–66.
6. Nishikawa, T.; Isobe, M. Synthesis of Tetrodotoxin, a Classic But Still Fascinating Natural Product. *Chem. Rec.* **2013,** *13* (3), 286–302.
7. Hinman, A.; Du Bois, J. A Stereoselective Synthesis of (−)-Tetrodotoxin. *J. Am. Chem. Soc.* **2003,** *125,* 11510–11511.
8. Ohyabu, N.; Nishikawa, T.; Isobe, M. First Asymmetric Total Synthesis of Tetrodotoxin. *J. Am. Chem. Soc.* **2003,** *125,* 8798–8805.
9. Nishikawa, T.; Urabe, D.; Isobe, M. An Efficient Total Synthesis of Optically Active Tetrodotoxin. *Angew. Chem. Int. Ed.* **2004,** *43,* 4782–4785.
10. Sato, K.-i.; Akai, S.; Sugita, N.; Ohsawa, T.; Kogure, T.; Shoji, H.; Yoshimura, J. Novel and Stereocontrolled Synthesis of (±)-Tetrodotoxin From *myo*-Inositol. *J. Org. Chem.* **2005,** *70* (19), 7496–7504.
11. Umezawa, T.; Hayashi, T.; Sakai, H.; Teramoto, H.; Yoshikawa, T.; Izumida, M.; Tamatani, Y.; Hirose, T.; Ohfune, Y.; Shinada, T. Total Synthesis of (−)-5, 6, 11-Trideoxytetrodotoxin and its 4-Epimer. *Org. Lett.* **2006,** *8,* 4971–4974.
12. Adachi, M.; Imazu, T.; Isobe, M.; Nishikawa, T. An Improved Synthesis of (−)-5,11-Dideoxytetrodotoxin. *J. Org. Chem.* **2013,** *78,* 1699–1705.
13. Xu, S.; Ciufolini, M. A. Formal Synthesis of (±)-Tetrodotoxin *via* the Oxidative Amidation of a Phenol: On the Structure of the Sato lactone. *Org. Lett.* **2015,** *17,* 2424–2427.
14. Maehara, T.; Motoyama, K.; Toma, T.; Yokoshima, S.; Fukuyama, T. Total Synthesis of (−)-Tetrodotoxin and 11-norTTX-6 (*R*)-ol. *Angew. Chem.* **2017,** *129,* 1571–1574.

# Abbreviations

| | |
|---|---|
| **(−)-DIPT** | Diisopropyl tartrate |
| **$(COD)_2RhCl$** | Cyclooctadiene rhodium chloride dimer |
| **$(DHQ)_2AQN$** | Hydroquinine anthraquinone-1,4-diyl diether |
| **$(DHQ)_2PHAL$** | Hydroquinine 1,4-phthalazinediyl diether |
| **$(dppf)PdCl_2$** | [1,1′-Bis(diphenylphosphino)ferrocene]-dichloropalladium(II) |
| **$(imd)_2C{=}S$** | Carbonyldiimidazole |
| **(*S*)-DTBM-SEGPHOS** | (*S*)-(+)-5,5′-Bis[di(3,5-*di-tert*-butyl-4-methoxyphenyl)phos phino]-4,4′-bi-1,3-benzodioxole |
| **$[Pd_2(dba_3)]$** | Tris(dibenzylideneacetone)dipalladium |
| **AcOH** | Acetic acid |
| **AIBN** | Azobisisobutyronitrile |
| **Ar** | Aryl |
| **AZADOL** | 2-Azaadamantane-*N*-oxyl |
| **BAIB** | Bis(acetoxy)iodobenzene |
| **BINOL** | 1,1′-Bi-2-naphthol |
| **Bn** | Benzyl |
| **BOP** | (Benzotriazol-1-yloxy)tris(dimethylamino)-phosphonium hexafluorophosphate |
| **BOPCl** | Bis(2-oxo-3-oxazolidinyl)phosphinic chloride |
| **CAN** | Ceric ammonium nitrate |
| **CbzCl** | Benzyl |
| **CbzCl** | Benzyl chloroformate |
| **$Cp_2Zr(H)Cl$** | Chloridobis(η5-cyclopentadienyl)-hydridozirconium |
| **$Cp_2ZrHCl$** | Bis(cyclopentadienyl)zirconium(IV) chloride hydride |
| **CSA** | Camphor sulfonic acid |
| **DABCO** | 1,4-Diazabicyclo[2.2.2]octane |
| **DBAD** | Di-*tert*-butyl azodicarboxylate |
| **DCC** | *N,N′*-dicyclohexylcarbodiimide |
| **DDQ** | 2,3-Dichloro-5,6-dicyano-1,4-benzoquinone |
| ***de*** | Diastereomeric excess |
| **DEF** | *N,N*-diethylformamide |
| **DIAD** | Diisopropyl azodicarboxylate |
| **DIBAL** | Diisobutylaluminum hydride |
| **DIEA** | Diisopropylethylamine |
| **DMAP** | 4-Dimethylaminopyridine |
| **DMDO** | Dimethyldioxirane |
| **DMF** | Dimethylformamide |
| **DMSO** | Dimethyl sulfoxide |
| **DPPA** | Diphenylphosphoryl azide |
| **EDCI** | *N*-Ethyl-*N′*-(3-dimethylaminopropyl)carbodiimide hydrochloride |
| **EEACE (enzyme)** | Electric eel acetylcholinesterase |

| | |
|---|---|
| **Fmoc-Cl** | 9-Fluorenylmethoxycarbonyl chloride |
| **HATU** | Hexafluorophosphate azabenzotriazole tetramethyl uronium or 1-[bis(dimethylamino)methylene]-1*H*-1,2,3-triazolo[4,5-*b*]pyridinium 3-oxide hexafluorophosphate |
| **HBTU** | 3-[*Bis*(dimethylamino)methyliumyl]-3*H*-benzotriazol-1-oxide hexafluorophosphate |
| **HMPA** | Hexamethylphosphoramide |
| **HOBt** | Hydroxybenzotriazole |
| **HPLC** | High-performance liquid chromatography |
| **IBX** | 2-Iodoxybenzoic acid |
| **KHMDS** | Potassium hexamethyldisilazide |
| **LDA** | Lithium diisopropylamide |
| **LHMDS** | Lithium hexamethyldisilazide |
| **$MeReO_3$** | Methylrhenium trioxide |
| **MOM** | Methoxymethyl acetal |
| **MOMCl** | Chloromethyl methyl ether |
| **NCS** | *N*-Chlorosuccinimide |
| **NDMBA** | 1,3-dimethylbarbituric acid |
| **NMO** | *N*-methylmorpholine-*N*-oxide |
| **Ns** | Nosyl (4-nitrobenzenesulfonyl) |
| **NsCl** | 4-Nitrobenzenesulfonyl chloride |
| **P(2-Fur)** | Disodium tetracarbonylferrate |
| **PDC** | Pyridinium dichromate |
| **$Ph_3P$** | Triphenylphosphine |
| **PhIO** | Iodosobenzene |
| **PMBBr** | 4-Methoxybenzyl bromide |
| **PMHS** | Polymethylhydrosiloxane |
| **PPA** | Polyphosphoric acid |
| **PPTS** | Pyridinium *p*-toluenesulfonate |
| **PTS** | *p*-Toluenesulfonic acid |
| **Red-Al** | Sodium bis(2-methoxyethoxy)aluminum hydride |
| **TBAB** | Tetra-n-butylammonium bromide |
| **TBAF** | Tetra-*n*-butylammonium fluoride |
| **TBDPS** | *tert*-Butyldiphenylsilyl |
| **TBS** | *tert*-Butyldimethylsilyl |
| **TBSOTf** | *tert*-Butyldimethylsilyl triflate |
| **TEMPO** | 2,2,6,6-Tetramethyl-1-piperidinyloxy |
| **$Tf_2O$** | Trifluoromethanesulfonic anhydride |
| **TFA** | Trifluoroacetic acid |
| **TFAA** | Trifluoroacetic anhydride |
| **TFEOH** | Trifluoroethanol |
| **TIPSOTf** | Triisopropylsilyl trifluoromethanesulfonate |
| **TMEDA** | Tetramethylethylenediamine |
| **TMS** | Trimethylsilyl |
| **TOlSH** | 4-Methylbenzenethiol |
| **TPAP** | Tetrapropylammonium perruthenate |
| **TPP** | Triphenylphosphine |
| **Ts** | Tosyl (4-methylbenzenesulfonyl) |
| **$VO(acac)_2$** | Vanadyl acetylacetonate |

# Index

*Note*: 'Page numbers followed by "f" indicate figures.'

Kolibri 360 GmbH
Ferdinand-Jühlke-Straße 7
99095 Erfurt, Deutschland
produktsicherheit@kolibri360.de